ROUTLEDGE LIBRARY EDTIONS: GLOBAL TRANSPORT PLANNING

Volume 2

CAR OWNERSHIP FORECASTING

CAR OWNERSHIP FORECASTING

E. W. ALLANSON

LONDON AND NEW YORK

First published in 1982 by Gordon and Breach, Science Publishers, Inc

This edition first published in 2021
by Routledge
2 Park Square, Milton Park, Abingdon, Oxon OX14 4RN

and by Routledge
605 Third Avenue, New York, NY 10017

Routledge is an imprint of the Taylor & Francis Group, an informa business

Copyright © 1982 by Taylor & Francis

All rights reserved. No part of this book may be reprinted or reproduced or utilised in any form or by any electronic, mechanical, or other means, now known or hereafter invented, including photocopying and recording, or in any information storage or retrieval system, without permission in writing from the publishers.

Trademark notice: Product or corporate names may be trademarks or registered trademarks, and are used only for identification and explanation without intent to infringe.

British Library Cataloguing in Publication Data
A catalogue record for this book is available from the British Library

ISBN: 978-0-367-69870-6 (Set)
ISBN: 978-0-367-74475-5 (Volume 2) (hbk)
ISBN: 978-0-367-74484-7 (Volume 2) (pbk)

Publisher's Note
The publisher has gone to great lengths to ensure the quality of this reprint but points out that some imperfections in the original copies may be apparent.

Disclaimer
The publisher has made every effort to trace copyright holders and would welcome correspondence from those they have been unable to trace.

CAR OWNERSHIP FORECASTING

E.W. Allanson

GORDON AND BREACH SCIENCE PUBLISHERS
London   New York   Paris

Copyright © 1982 by Gordon and Breach, Science Publishers, Inc.

Gordon and Breach, Science Publishers, Inc.
One Park Avenue
New York NY 10016

Gordon and Breach Science Publishers Ltd.
42 William IV Street
London WC2N 4DE

Gordon & Breach
58 rue Lhomond
75005 Paris

Library of Congress Cataloging in Publication Data

Allanson, E.W., 1948–
   Car ownership forecasting.

   (Transportation studies, ISSN 0278–3819; v. 1)
   Includes bibliographies and index.
   1. Automobile ownership—Forecasting. 2. Prediction theory. I. Title. II. Series.
HE5613.A73 1982   388.3'422   81–13411
ISBN 0–677–05690–7   AACR2

Library of Congress catalog card number 81-13411. ISBN 0 677 05690 7. ISSN 0278–3819. All rights reserved. No part of this book may be reproduced or utilized in any form or by any means, electronic or mechanical, including photocopying, recording, or by any information storage or retrieval system, without prior permission in writing from the publishers. Printed in the United Kingdom.

Contents

Series Editors' Introduction	vii
Preface	viii
Acknowledgements	x
Abbreviations	xii
Introduction	1
1 Modelling in the UK 1900–1960	9
2 The Road Research Laboratory Model and its Development to 1973	17
3 Econometric Modelling and the Problem of Disaggregation	29
4 The Impact of Government Intervention	44
5 A Review of US Modelling Prior to 1962	80
6 Recent Developments in the USA and Australia	94
7 National Theory: Local Practice	122
Conclusion	135
Additional References	141
Appendix 1: Summary of the TRRL Logistic Curve Procedure	143
Appendix 2: Population Projection Methods used in the UK	146
Appendix 3: Car Ownership in the USA and UK 1900–1979	149
Index	151

Series Editors' Introduction

This particular work is the first of a broad-ranging series of books which will cover many of the varied aspects of transportation. The subject area will be generally divided into two parts: the first, dealing with planning and technological aspects of transportation, the second with specialized transportation.

Within the context of this series, technology and planning will include the wide spectrum of various aspects of the design and planning of vehicles and infrastructure for the transport of freight and passengers, as well as operational and management considerations. The general aim of the planning and technology series is to provide readers with the state of the art and to summarize the status of transportation.

The second part of the series will seek to generate monographs dealing with improving the mobility of those groups in society increasingly characterised as the transportation disadvantaged, particularly, but not exclusively, the elderly, the disabled and families with low income. It is anticipated that the content of these books will be derived largely from research, policy analysis and documental field experience. The subject matter will include advances in relevant technology, service and methods demonstrations, improved planning and methodology, major or proposed changes in public policy and innovative proposals for system development or change.

Occasionally, more specific monographs will be published, presenting the results of individual studies into areas of special interest to planners and technologists.

As with any monograph series, the emphasis is on currency of information, and the material will be of interest to the transportation practitioner, the postgraduate student and academics working in the field.

NORMAN ASHFORD
WILLIAM G. BELL

Preface

The object of this book is to provide a straightforward, readable commentary on the development and performance of prediction procedures. It is a topic which continues to generate a large amount of research and discussion; a problem which may appear at first sight subject to precise understanding and explanation, but which has aroused considerable technical and philosophical scepticism.

In content the book is primarily concerned with technical problems which have been experienced by modellers and forecasters. It is hoped that the discussion provides a basic appreciation of the key points, whether they are of a mathematical nature or otherwise. Throughout the book there is a theme which attempts to relate the academic debate surrounding the issue to technical, rather than philosophical, concepts. One of the main aims is to develop an appreciation of the way in which forecasters have tried to come to terms with the uncertain economic and social climate of the last quarter of the twentieth century.

The book is structured on a chronological as well as geographical basis. This has been possible because of the particular way in which research car ownership has been organised. Exchanges between authorities in different western countries has been very limited. Consequently, the book has been divided into two major sections. Firstly, a general description of the evolution of various approaches is discussed in the context of work conducted in the United Kingdom. This is followed by a comparative study of the development of procedures in the United States and, briefly, Australia. The book concludes with an assessment of the problems associated with the application of techniques by regional and local planning authorities in the United Kingdom.

A full rendition of every example of model development and usage is beyond the scope of this work. Subsequently, direct reference has been afforded to instances where significant developments or valid criticisms have been made. Most chapters contain a select list of references which have been cited in the text, whilst a (not exhaustive) bibliography provides scope for further reading as well as an indication of the scale and range of contributions in the field.

E.W. ALLANSON

Acknowledgements

This book represents the development of a postgraduate thesis originally submitted at the University of Salford, England. Particular thanks are due to Mr Roy Thomas of the Department of Civil Engineering, University of Salford, for his supervision of the original thesis. Professor Norman Ashford, of the Department of Transport Technology, University of Loughborough, offered valuable advice on the final development of the theme, as well as constructive comments on the manuscript. Additional assistance has been gratefully received from Professor Moshe Ben-Akiva, Department of Civil Engineering, Massachusetts Institute of Technology and Mr Alan Wann, formerly Group Engineer, Cleveland County Council. Helen Allanson typed, and re-typed, the manuscript whilst Elizabeth Richardson provided valuable advice on grammatical matters.

A review of this nature must, of necessity, make reference to a wide range of published material. Thanks are due to the following concerns for granting permission to either quote passages, or include diagrams, from books or reports published by them; the Bodley Head, Faber and Faber, Her Majesty's Stationery Office, the Department of Motor Vehicles (State of California), the United States Department of Transportation, Prentice-Hall Publishers Inc., the Eno Foundation for Transportation, the Publication Division of Massachusetts Institute of Technology, Diana Crawfurd Ltd, The Transport and Road Research Laboratory, the Transportation Research Board, Cambridge Systematics Inc., the New South Wales Ministry of Transport and the MIT Press.

In addition, a number of publishers have granted permission to reproduce extracts from articles which have appeared in the following journals; *Journal of Transport Economics and Policy*, *Traffic Engineering and Control*, *Roads and Road Construction* (now *Highways and Public Works*), *Econometrica*, *Scottish Journal of Political Eco-*

nomy, Transportation, American Economic Review, Canadian Journal of Economics, Urban Studies. Mr H.R. Kirby, Mr K.J. Button and Professor M.E. Ben-Akiva kindly agreed to the use of material contained in unpublished papers.

In order to improve clarity of presentation it has been necessary to re-draft several of the diagrams extracted from publications. Any uncredited views expressed in the book are those of the author and do not necessarily conform with those of his present or past employers.

Abbreviations

ACTRA	Advisory Committee on Trunk Road Assessment
CATS	Chicago Area Transportation Study
DOE	Department of the Environment (UK)
DTp	Department of Transport (UK)
FAPS	Future Automobile Population Stochastic Model
FES	Family Expenditure Survey
GDP	Gross Domestic Product
GLC	Greater London Council
NSW	New South Wales
NTS	National Travel Survey
OPCS	Office of Population, Censuses and Surveys (UK)
RHTM	Regional Highway Traffic Model
RRL	Road Research Laboratory (TRRL from 1972 onwards)
SATS	Sydney Area Transportation Study
SMMT	Society of Motor Manufacturers and Traders
TPA	Transportation Planning Associates
TPP	Transport Policies and Programme
TRRL	Transport and Road Research Laboratory (RRL prior to 1972)

Introduction

The Scope of the Problem

THIS COMMENTARY is concerned with a critical assessment of car ownership prediction techniques as they have been developed in the United Kingdom, the United States and other developed countries since 1950. A great deal of work has already been carried out in the field of car ownership prediction and, owing to a combination of circumstances, the degree of research and publication of results has tended to accelerate in size and scope in recent years. Since much of the earlier work has become generally accepted as valid or, alternatively, superseded, this study concentrates on the developments of the last decade. Not only are the wider theoretical issues considered, but an attempt is made to examine the practicalities of the application of car ownership prediction at national, regional, local and individual (or household) level, with particular reference to the work of several (typical) local authorities.

For many years now car ownership prediction methods have been a subject surrounded by controversy. This has resulted in a lively and continuing discussion in the political arena at various levels as well as in the academic and general press. Arguments are presented at a variety of levels and from a number of standpoints. Often, car ownership prediction techniques are introduced into discussions which are concerned with wider issues, including the use of scientific quantitative philosophy in the field of economic, social and transport planning. This may be regarded as one extreme, although it is felt inappropriate to include a long essay on the validity of the use of scientific (or, at least, mathematical) models in social and economic planning. Further discussion on the subject could add little to the mountain of literature already available and it must be admitted that, in a world composed of a large number of competitive nation states, many of which are highly industrialised and which contain sophisticated economic and social systems, there is no realistic alternative to some level of application of logical scientific reasoning to the problems of planning for the economic, social and transport conditions of the foreseeable future.

At the other extreme is the over-detailed analysis and application of prediction techniques. A phenomenon which involves such a notable human input and which is influenced by a wide range of variables cannot be treated with the same level of simple analysis as is applied to many socio-economic variables (despite the long history of the application of simple extrapolatory, but not explanatory, models). The result has been a magnitude and depth of research activity into the field which is indicated by the (not exhaustive) accompanying bibliography. Research and publication has accelerated since 1970, to the extent that the production of papers has taken on the form of the predictive curve used by one of the major authorities in the field.

Growth in car ownership

The growth in automobile ownership in the United States has often been used as an empirical base for the development of models for use in other countries (see Appendix 3). In spite of a continuing search for evidence of "saturation" in the number of vehicles in use, the growth in ownership has followed a steady path since 1945. Previous reductions in the growth rate are explained in terms of the effects of the Depression of the 1930's and the impact of the Second World War.

In national terms several notable points on the car ownership curve have been approached and then passed. By 1922 the rate of ownership of vehicles to persons already exceeded 0.1. Seven years later, just before the Depression began, the rate stood at 0.19, a level not experienced in many European countries until relatively recent times. By 1950 the renewed trend had already taken hold and there was very little deviation in growth rate until 1977, by which time 99,904,000 automobiles were in use (representing an ownership rate of 0.46 cars per person).

Ownership in the UK and most European countries has lagged far behind that experienced in the USA (see Appendix 3). Here the general availability of private vehicles has been a phenomenon of the period since World War II. To a much greater extent than in the

United States, ownership was restricted to the affluent classes until relatively recently. Even as late as 1964, at a time when the construction of a national motorway network had already commenced, 62 per cent of UK households were without access to private vehicles. It is worthy of note that the rate of ownership experienced in the US by 1929 was not achieved in the UK until 1966. This gives some indication of the contrasting role which the automobile has played in these countries and the extent of the time lag between the two. Despite a relatively slow start and the effects of economic depression in recent years, the UK is now a considerable way down the road to emulating the pattern already experienced in the US and several other expansive western economies. Most forecasters agree that such trends will continue; the major problems have been associated with the future rate of growth and the point at which new car registrations will represent only the replacement of scrapped vehicles.

The importance of car ownership forecasting

The critical role played by the private automobile in the social and economic life of the USA is easily appreciated. Consistent levels of ownership and use have had their influence upon all forms of social and economic activity, and on the resultant pattern of land use and civic design. To a much greater extent than in most countries, urban development and car ownership have been inextricably linked.

In Britain, at least on a superficial level, the effects of the "mixed blessing" of car ownership have been more problematical. Despite the fact that more than 60 per cent of British households enjoy access to at least one private car, a long national debate continues on the moral, social, and economic consequences of the move towards those levels of car ownership already experienced in the United States.

Even in the UK there are few aspects of life (and even death) which have not been affected by the growth in car ownership. A large number of agencies are directly interested in the future patterns of ownership and use, and, to a large number of others, the problem is of more than academic interest.

In many countries central government has a particular interest in forecasts of car ownership, as a transport authority whose responsibility it is to provide the necessary physical infrastructure to accommodate the demand for road space. Car owners and users are a major source of government revenue. In the UK, in 1980–81, the Government expected to extract more than £7,000 million from car users by way of sales tax on cars, petrol, parts, value-added tax on labour, and so on. Clearly the degree of consumer and commercial expenditure on cars, coupled with levels of taxation, has a direct influence on capacity for spending on other items. This is particularly important, because national expenditure on motor cars has consistently accounted for a greater share of consumer expenditure as more and more households have entered the single and multiple car ownership categories. A household which becomes car owning for the first time immediately switches a substantial proportion of household resources into provision of personal "transport", even though the physical expression of this change in status may spend all but three per cent of its life in a stationary condition.

Paradoxically, increases in car ownership can lead to a reduction in overall accessibility for the community as a whole. In a car oriented society alternative forms of transport usually decline in quality. There is a significant causal link between car ownership levels and ridership on public transport, especially in non-metropolitan areas. Car ownership forecasts are a major consideration in assessing the future viability and role of public transport, especially in a situation in which quality of access is viewed on a strict commercial, rather than social, basis. Many countries are now committed in political, social and commercial terms to encouragement of, and accommodation for, mass car ownership and use. In most western countries more than half the electorate are members of car owning households. Consequently, the possibility of political acceptance of a doctrine aimed at slowing down the move towards mass car ownership reduces in line with the increase in vehicles.

At a national, strategic level, future rates of car ownership and use are of critical importance. There is a particularly vivid perception of the importance of the availability and price of energy. The popular belief is that transport, in particular private road transport, represents a wasteful use of energy resources. It is inappropriate to discuss this at length here, only to note that total consumption of energy in transport has risen in line with car ownership, as a consequence of the more expansive life style which car use has permitted. In 1979, transport accounted for 16 per cent of energy consumption in the UK. Of this, road transport used more than three quarters of the total. The proportion of total energy consumption given over to transport has increased consistently since the early 1960's as the impact of mass car ownership, coupled with greater affluence, has continued to influence the land use/movement system.

At the regional, state and local level, car ownership rates are a significant consideration in the overall social and economic planning system. The layout of residential and industrial areas, rate of provision of garaging and parking space, and the size and spacing of shops, offices and services tend to be conditioned by present and future rates of ownership, and, of course, changes in traffic levels play a significant part in the justification for and design of new roads. The problem is thrown into relief by consideration of roads in residential and commercial areas set out before the early 1960's, and the degree of blight which has been caused by superimposition of a car based society upon them. Added to this has been the effect which greater car ownership has had on the pattern of retailing, and the negative consequences for those establishments which have been unable, for one reason or another, to adapt to changing patterns of access.

Of course, a great number of people are directly employed in the car industry and motor trade, or in ancillary concerns. Future levels of demand for cars are of great importance to the industry. Forecasting the future market for cars, short and long term, has been a central research theme, particularly in the United States. The problem is particularly important in the context of the approach towards mass car ownership and the resultant decline in "first time" purchase.

Types of model

In spite of a relatively long history of forecasting and modelling car ownership it has usually been possible to classify most procedures into one of a small number of categories. It should be noted here that a number of models have been purely descriptive or analytical of historic data and there is no evidence that they have ever been used for predictive purposes. Nevertheless, predictive or otherwise, a restricted number of approaches have been used.

A number of commentators have sought to produce taxonomies of scientific forecasting in general. In the field of car ownership there lies a broad distinction between those models which involve an extrapolation of an existing trend, evident from a set of historic data, compared to alternative "causal" models. Extrapolation models have been termed "naive", not in the sense of being simple, (many are complex in statistical terms), but because they do not lend themselves to explanation of the underlying causes of change. Their use and development usually involves an assumption that the general social and economic environment will not change significantly during the forecasting period. This has been regarded as a weakness by many commentators and has been used as a primary justification for experimentation with a wide range of alternative procedures.

Most of these alternative models fall into the causal category. This has usually involved an examination of the degree of explanation of change in the dependent variable (car ownership) in terms of one or more independent variables. Although popularly regarded as more robust than extrapolation, these procedures still involve some form of independent forecast of the pattern of change in the independent variables. Several recent observers have noted the greater margin of error which is involved in causal forecasting, in an uncertain world, and have suggested that the most valid procedure should be some combination of the two methods.

A further classification which is possible is based on the level of aggregation of the data used in the development and application of models. This has formed the basis of a long academic debate on the merits associated with the use of local data in preference to national and regional figures. There has been particular concern voiced over the use of statistical means and averages derived from large data sets, and the resultant tendency to conceal significant variations within the zonal unit. Partly as a response to this we have seen the more recent development of household-based causal models in which the data and methodology is "disaggregated" down to the level of the basic behavioural unit, the household. New sources of error associated with forecasting household behaviour have resulted in the introduction of stochastic procedures, involving the application of probability theory. At the time of writing, a lively debate is still in progress on the relative merits of all of these methods, in the US, the UK and other countries.

Two features of modern work are noteworthy in the context of extremes in approach to the problem. Firstly, there has been a distinct tendency to introduce new ideas at the expense of reduced credibility and statistical reliability. Future prediction of any socio-economic variable should be a highly qualified activity and it is noteworthy that despite frequent references to the dangers of blind acceptance of "official" figures there still exists a need and desire on the part of decision makers at national and local level to carry on in such a manner. This is in spite of the recent experiences of western economies since the events of late 1973 upset the "balanced" rate of economic development upon which so many socio-economic growth predictions were based. Despite some apparent success in the past fifteen years, car ownership and car use prediction remains an activity subject to much wider margins of error than many other more rationally explained areas of consumer expenditure. Early attempts to substitute car purchase for washing machines or television sets in consumer purchase models soon exposed aspects of irrationality (or extra-variate explanation) which suggested that the car and its purchase, ownership, and use demanded special attention.

A second feature of the car ownership prediction "discussion" is the tendency for *a priori* opinions and hypotheses to influence the approach and techniques of researchers. Content of papers and reports and subsequent discussion in the form of letters, papers, and seminars has exposed fixation in the attitude of some of the participants. One of the prime requirements of scientific research is impartial objectivity. In the car ownership prediction field badly fitting data and unexpected variations have sometimes been discarded or omitted in the interests of statistical significance. Extensive and complex calibration procedures often assume greater importance than the original problems of car ownership prediction while unexplained variations are often noted but equally often are left to "others" to analyse in detail.

There have developed clear factions in the car ownership prediction field with the most prominent composed of those whose academic training appears to have been in statistics (and mathematics) on the one hand and economics on the other. Other individuals or groups are drawn into the discussion but the sociologists and psychologists, who should be playing their part, are conspicuous by their absence.

Factors which influence car ownership

It requires little imagination to appreciate the type of variables which are likely to influence the future pattern of car ownerhsip. Several contributors have sought to provide an index of factors which have been shown to have some significant effect. Although open to criticism regarding arrangement and inclusion of certain variables, Figure 1·1 provides a useful summary of the magnitude and scope of the problem. To provide a logical explanation of patterns of ownership in terms of the many variables shown has become the central theme of modern work, moving away from relatively simple extrapolation of past and present trends.

In summary, there is evidence of continuing academic debate and factionalising. At the one extreme is the "why bother predicting?" argument which is not the concern of this work. At the other extreme we have examples of criticism of, say, techniques used to calibrate innumerable alternative parameter values in a multiple-parametered extrapolative car ownership prediction curve equation (when there is a need for much more relevant discussion of causal relationships).

1 Modelling in the UK 1900–1960

DESPITE AN increasing research emphasis on traffic prediction during the past twenty years, it would be incorrect to assume that the issue was totally ignored during earlier periods. Although the amount of published material is small there is evidence that British Governments and other bodies have at least been aware of the existence of a growing private vehicle population for most of the present century. It was, however, not until the 1920's (in the aftermath of the First World War) that the growing number and use of private vehicles required noteworthy changes in highway infrastructure and a need to indulge in some limited prediction of increase in numbers with subsequent "planning" proposals and strategy.

Plowden (1971) has provided a limited summary of the official central government philosophy and policy adopted during the interwar period. Surprisingly, there do appear to have been some limited scientifically-reasoned forecasting procedures along similar lines to those later adopted in the new conditions of the 1960's. In addition, the inter-war period provoked just the degree of temporal and geographical variations in economic circumstance to test some of the broader assumptions concerning the influence of income on car ownership. As might be expected, the immediate post-war growth in new registrations had declined after 1925 before reviving again from 1931 onwards. Despite severe economic depression in certain regions expansion in the south, coupled with considerable real reductions in car and petrol (running) prices, was sufficient to maintain an increasing growth rate.

Without any apparently rigorous statistical analysis, Plowden reached the conclusion that a major reason for the continuing demand for cars was the steady reduction in cost of purchase and operation. With the growing use of production line techniques, the cost of new cars fell by more than 50 per cent between 1920 and 1928. Prices in 1938 were 33 per cent less than they had been ten years earlier. In line

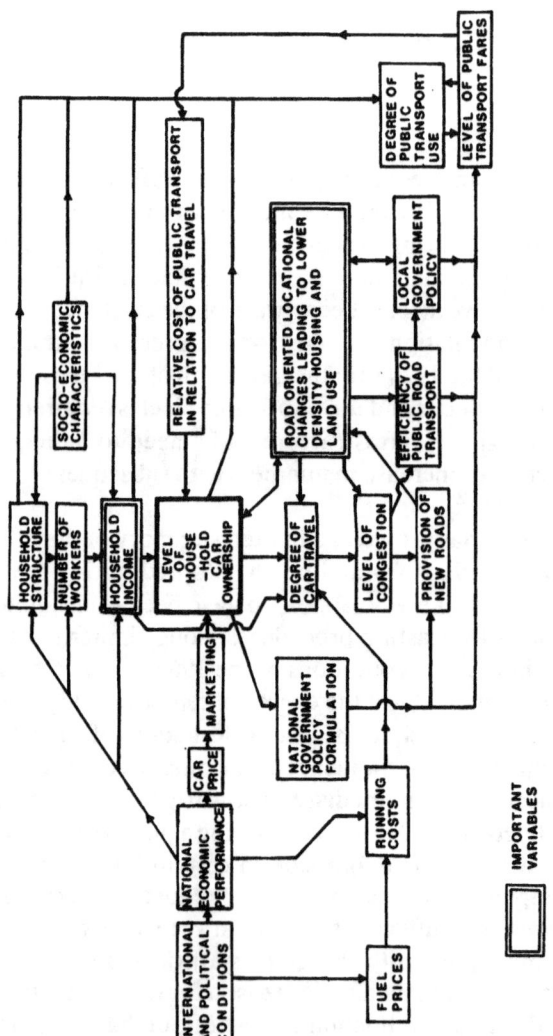

FIGURE 1.1. Factors influencing the level of car ownership.

with new prices, running costs also fell. The combined cost of purchase and use fell more than any other item of household expenditure.

To a much greater degree than exists in modern times, and refuting some of the logic applied to present day patterns, there was a notable regional variation in ownership levels as is shown by Figure 1·2. Under such conditions, any attempt at national forecasting on the basis of aggregation of data was liable to misuse, though there is little evidence that central governments expended very much energy on road and traffic matters, other than in the area of road safety, until the changing pattern of car ownership and use became evident in the late 1950s. What limited forecasting there was soon proved to be widely inaccurate and reflected the rule-of-thumb philosophy upon which it was based.

In 1929 (with annual registrations still falling but with nearly a million cars already in use) the SMMT [Society of Motor Manufacturers and Traders] continuously revised their estimate of saturation point to a million and a quarter (from an earlier 1927 estimate of one million). From 1932 sales began to pick up again at an ever-increasing rate. *The Economist* expressed astonishment at the resistance of the motor industry to the depression. There were a million and a quarter cars — "saturation point" — on British

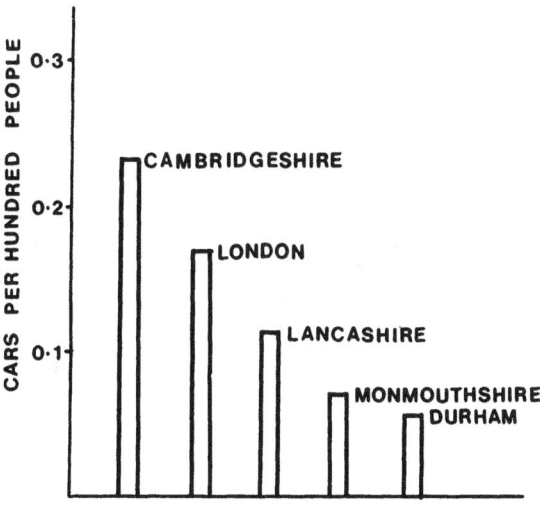

FIGURE 1·2. Variation in car ownership in selected areas in 1927. (Source: Ministry of Transport Statistics.)

roads by 1934; by the time the war began there were over two million. (Plowden, 1971, p. 264.)

It is clear that there were problems in estimating that parameter which has been a continual source of debate over the past twenty years, namely saturation point, or that level of ownership which reflects a stable ratio of cars per head of population with no latent desire for car ownership on the part of any group or individual within that population. The situation in the 1930's provides an interesting comparison with more recent assumptions on saturation levels. Pre-war commentators obviously assumed that car ownership was clearly beyond the means or desires of the working-class section of the community. In recent years, it has been assumed that the vast majority of the population desire car ownership or at least free access to private vehicle use, and that the final constraints are financial or related to age and physical ability to drive. Clearly, the popular ideas of the 1930's suggest that any assumption regarding desire for car ownership should be open to modification in the light of experience.

Although the nature and scale of the problem differed considerably from that which faced transport planners in the 1960's, it is clear that (in the 1930's) there already existed some of the pitfalls of prediction based on apparently rational variables and relationships. The parallel between the inter-war period and the post-1973 economic "depression" is worthy of comment. In so far as a large faction of predictive researchers base their hypotheses on economic variables, any logical observation of the inter-war period, using economic data aggregated at national level, would have predicted a stagnant level of car ownership and use. In reality, spatial and social variations in income and purchasing power, coupled with other social and economic variables, resulted in a relatively high rate of increase in vehicle ownership (among the middle class) during the 1930's.

The (already) well-established tendency to under-predict the growth of vehicle ownership was evident once again during the 1950's, only now the "problem" was much more serious as ownership of motor vehicles began to extend beyond the bounds of upper and middle classes. It is clear that prediction procedures used in the 1930's were based on a loosely accepted hypothesis of the relationship between ownership and income distribution and that although it is accepted that such a relationship does exist, ownership rose much faster than real incomes for a number of reasons. During the early 1950's it was estimated that at least 50 per cent of vehicles were owned or partly financed by firms or business interests. Coupled to this was the influence of widespread use of hire purchase which predictions had not considered. So a familiar tendency to underestimate growth rates and saturation levels is evident during the period 1945–1960, forming only one aspect of a limited governmental involvement in the road transport problem. Contemporary predictions throw into relief the relative success which has been achieved on the basis of work carried out since 1962 at the Road Research Laboratory (RRL).

In 1945 the Ministry [of Transport] estimated that traffic increases by 1965 would be 75 per cent over 1933 figures in urban areas and 45 per cent more in rural areas; but by 1950 — five years later — the number of vehicles had already nearly doubled compared with 1933. In 1954 the Ministry told highway authorities to plan for a 75 per cent increase in traffic by 1974. But by 1962 the number of vehicles was already 75 per cent above the 1954 figure. In 1957 the Ministry forecast that there would be eight million vehicles by 1960; but there were 9 million. In 1959, the Ministry of Transport forecast that there would be 12½ million vehicles by 1969, but this figure was passed in 1964. (Tetlow and Goss, 1968, p. 83.)

Other government policies on transport, and road provision in particular, reflected the assumptions inherent in these predictions.

Views about the relationship of motoring to both class and income — and probably also the residual notion that car ownership is a luxury of which the state, in providing for the necessities of life, need take no account — were built into the provision of garage space and of roads in *public* housing developments. (Plowden, p. 327.)

As far as road construction is concerned, the story of inactivity and piecemeal investment during 1945–58 is well known and needs no expanded treatment. Legal powers to build motorways existed under the Special Roads Act of 1949 but it was not until December 1958 that the first section of motorway was opened.

The need for accurate and more dependable prediction led to new pressure and research interest with the first consideration of overseas (particularly USA) experience. The justification for accurate and soundly based prediction has not changed, only become more valid, as pressure on space and resources has increased to the extent that the rule-of-thumb techniques of the 1950's are totally inadequate. The vast amount of academic research activity of a statistical and economic nature may be justified solely on the grounds of reducing the probability of a slight percentage error, perhaps saving millions of pounds' unnecessary expenditure.

The 1950's also witnessed greater involvement in forecasting on the part of staff of the United Kingdom Road Research Laboratory. Although primarily concerned with growth rates in overall vehicular traffic some coverage was afforded to the specific problem of explanation of car ownership notably by Rudd (1951) and Reynolds (1954). These commentators gave only approximate estimations of future levels of ownership. More detailed projections were produced by Chandler (1958). Using a number of methods, including extrapolation, he concluded that, by 1968, the number of vehicles in use would reach more than ten million, possibly twelve million (the actual GB 1968 figure was 14.4 million).

The problem of provision of scientifically derived forecasts also received attention from Tanner (1958). On the basis of an analysis of growth patterns in the USA and parts of the UK, he noted that areas with higher levels of ownership tended to exhibit lower rates of growth, indicating evidence of a declining demand for first cars as a situation of "saturation" approached. Several linear extrapolations of this trend produced tentative "saturation levels" which would be achieved when the rate of growth fell to zero. Tanner postulated that growth in the UK could be expected to follow a curve exhibiting logistic characteristics, its shape being controlled by three parameters — base year rate of ownership, base year growth rate, and saturation level. Using a 1957 base this procedure was used to predict a 1975 forecast of 16.4 million vehicles (this compares closely with the actual 1975 value of 17.5 million). (Full details of the Tanner logistic procedure are discussed in Chapter 2.)

Within the United States, however, several authorities had already set about more detailed examinations of growth trends in car ownership and other consumer durables and it was in the latter field that much UK work was based prior to 1960. Some of this work pre-dated the Second World War, notably that of de Wolff (1938).

De Wolff used a time-series extrapolation of demand for new cars based on patterns of scrapping and replacement during the period

1921–1934. It was unfortunate that the time period chosen (available) coincided with economic conditions which were not typical of the post-war period, making the analysis largely invalid as a predictive exercise, but his use of a logistic growth curve for car purchase established a precedent in this field which was followed very vigorously by the Transport and Road Research Laboratory (UK) more than twenty years later.

In the immediate post-war period, a limited amount of research was carried out on the problem, mainly in the United States. There was a distinct tendency to look for economic explanations of existing trends. Typical is the work of Farrell (1954), who paid particular attention to economic explanation for changes in demand for cars. In classifying motor cars in the same category as other consumable durables, Farrell was guilty of an invalid assumption exposed by later researchers (notably Cramer, 1962) and in any case his work was largely concerned with an historical examination of trends in the USA between 1941 and 1953. He made no attempt to make any prediction of future rates of vehicle purchase and levels of ownership. Indeed Dawson, in the "discussion" which followed the presentation of Farrell's paper, noted the difficulties resulting from regarding motor cars as typical consumer durables. Not only was the concept of a saturation level less acceptable, but, already by the mid 1950's, the market in second-hand vehicles was of a size quite incomparable with any other durable such as washing machines or vacuum cleaners. This feature of the motor vehicle market has remained a notable constraint on analogous studies.

During the same period, Cramer was working on a similar area of study in the UK which produced a general economic model for ownership of consumer durables. This took into account household size and type of residential area as the major determinants of ownership of consumer durables. The study had already made an *a priori* assumption that in fact the nature of demand for motor cars was inherently different from that for other goods, on account of their special qualities and the non-economic benefits ownership conferred. So any attempt to compare growth rates and distribution patterns for cars on the one hand and television sets (for example) on the other led to inconclusive results. In the context of later work, the most notable development was the attempt to fit a sigmoid curve to the pattern of growth in ownership of consumer durables on the assumption that growth rates will begin to decline and ultimately fall to zero as the number of potential new owners is reduced (even allowing for moderate increases in population). Cramer used a modified version of the sigmoid curve, the lognormal approximation, in his applications to four durables (motor cars, television sets, washing machines, refrigerators). Work by Duesenberry (1949) on rates of diffusion had already contained the assumption that, given a normal income distribution, the rate of diffusion of a durable should follow such a logistic or sigmoid curve.

REFERENCES

Chandler, K.N., Traffic trends, Research Note RN/3174, Road Research Laboratory, Harmondsworth, 1958.
Cramer, J.S., *The Ownership of Major Consumer Durables*, Cambridge University Press, Cambridge, 1962.
Duesenberry, J.S., *Income, Saving and Theory of Consumer Expenditure*, Harvard University Press, Boston, 1949.
Farrell, M.J., The demand for motor cars in the United States, *Journal of the Royal Statistical Society* (A), 1954, pp. 171-193.
Plowden, W., *The Motor Car and Politics in Britain*, The Bodley Head, London, 1971.
Reynolds, D.J., An estimate of the demand for private cars in Great Britain in 1952, Research Note RN/2134, Road Research Laboratory, Harmondsworth, 1954.
Rudd, E., The relationship between the national income and vehicle registrations, Research Note RN/1631, Road Research Laboratory, Harmondsworth, 1951.
Tanner, J.C., An analysis of increases in motor vehicles in Great Britain and the United States, Research Note RN/3340, Road Research Laboratory, Harmondsworth, 1958.
Tetlow, J., and Goss, A., *Homes, Towns and Traffic*, Faber and Faber, London, 1968.
de Wolff, P., The demand for passenger cars in the United States, *Econometrica*, April, 1938, pp. 113-129.

2 The Road Research Laboratory Model and its Development to 1973

WHILST WORK on projections of consumer durable purchase in general was of no notable urgency during the period in question, growth in ownership of motor vehicles during the late 1950's and its consequences for the entire socio-economic structure of western society led to this particular phenomenon receiving special attention from central governments in the UK and elsewhere. Detailed analysis of the political and economic forces in play at the time and their implications on forecasting are beyond the scope of this work, but two major developments took place in the early 1960's which are still very relevant to the whole field of car ownership prediction. The first was the initiation of in-depth research at the Road Research Laboratory under Tanner and the second was the publication of the Buchanan Report on *Traffic in Towns* in 1963. The latter was primarily concerned with the urban traffic problem (as opposed to rural and inter-urban movement) but it did contain substantial sections on likely levels of car ownership and use which produced reasoned criticisms and had the effect of triggering off new efforts at explanation.

Official publication of the underlying philosophy and results of Tanner's work on vehicle number prediction took the form of a paper in *Roads and Road Construction* in 1962. Since, in effect, the basic methodology has changed little during the past 19 years — in spite of a number of further publications by Tanner and associates — there is some need to outline the original method, results and conclusions (and conditional qualifications). Tanner's original predictions have been updated on a number of occasions since and have been subjected to limited disaggregation (in the form of county predictions — Herrmann,

1968). It should also be noted that later discussion will be concerned with Tanner's more recent reports on the matter (1977, 1979) which represent a notable divergence from earlier (relatively inflexible) methodology.

Tanner has been noted for his preference in using time-series extrapolation over the alternative economic cross-sectional approach. Although we are given no clear positive indication as to why he opted for the use of a logistic extrapolation, he may have been influenced by earlier work in the USA (noted above). However, his reasons for not employing economic reasoning in place of trend analysis were that ownership is only partly explained by income and that, even if income was the main factor, it was not clear how this, in turn, should be extrapolated. In addition it was clear that if incvome forecasts were merely extrapolations of past trends it was preferable to omit this complication altogether and forecast directly from past trends in car ownership.

It is important to note the justifications for the use of the extrapolatory approach because they have remained central to the TRRL philosophy ever since, despite a consistent barrage of criticism from econometricians and others which has intensified during the past decade.

In addition to his negative reasoning on the relatively limited consideration of economic variables, Tanner expounded on the qualities of the use of logistic extrapolatory curve. He believed that it was reasonable to assume that past trends (the "known" part of the curve) would be smooth and that immediate growth rates would be close to those experienced in the few years prior to the base year (for prediction). He noted that several forms of curve were available, although, since a straightforward linear projection was not appropriate, there would have to be three controls or constant parameters on the curve. As other researchers had done before, he noted that the logistic curve could be justifiably applied here, with its three constants (one of which was an upper limit estimated as the result of an independent calculation). The upper limit would represent the maximum rate of ownership of cars/vehicles that would be likely to be expected. Thus three parameters were required as follows:

a) car ownership per person at base year

THE ROAD RESEARCH LABORATORY MODEL 19

b) rate of growth at base year
c) saturation level

Details on the means of fitting the logistic curve are provided in Appendix 1. What is more relevant to the discussion which has surrounded Tanner's work up to 1977 is the means by which the all-important saturation level is arrived at.

This too was explained in Tanner's 1962 paper and although modifications have since been introduced the basic choice of method and limiting variables remain unchanged. Although paying some (limited) note to the influence of income changes on vehicle ownership rates, most attention was placed on an examination of the apparent relationship between annual growth rate (percentage increase in ownership per year) and actual ownership (cars per head) over several time periods. Both sets of data were at county level (for UK) or state level (for USA), thus introducing a third spatial variable into the analysis, the problems of which were outlined (by Adams) more than ten years later. In the Tanner method, a linear regression line was estimated with several sets of data for the UK and USA for different time periods, and was found to slope negatively (indicating a reduction and eventual disappearance of growth at some time in the future). It was assumed that the intercept of the regression line on the x axis (cars per head) would represent no-growth and therefore (assuming only replacement of scrapped vehicles from that time hence) the general saturation level for the country (cars per head). Insertion of the parameter (saturation) into the logistic curve equation would therefore provide a pattern of growth up to saturation year, the latter being also derived from the curve fitting procedure. A graphical representation of the relationship between growth rate and level of ownership, used to forecast a saturation level, is shown in Figure 2·1. Figure 2·2 shows a typical logistic car ownership forecasting curve of the type employed in early RRL studies, complete with alternative saturation levels which have been calculated independently.

In addition to noting the "secondary" influence of income per head as a determinant of car ownership, Tanner also examined the influence of variations in residential/population density as an influence upon rates of car ownership. This variable has remained a prominent feature of all Tanner's published work on the subject, though there has

20 CAR OWNERSHIP FORECASTING

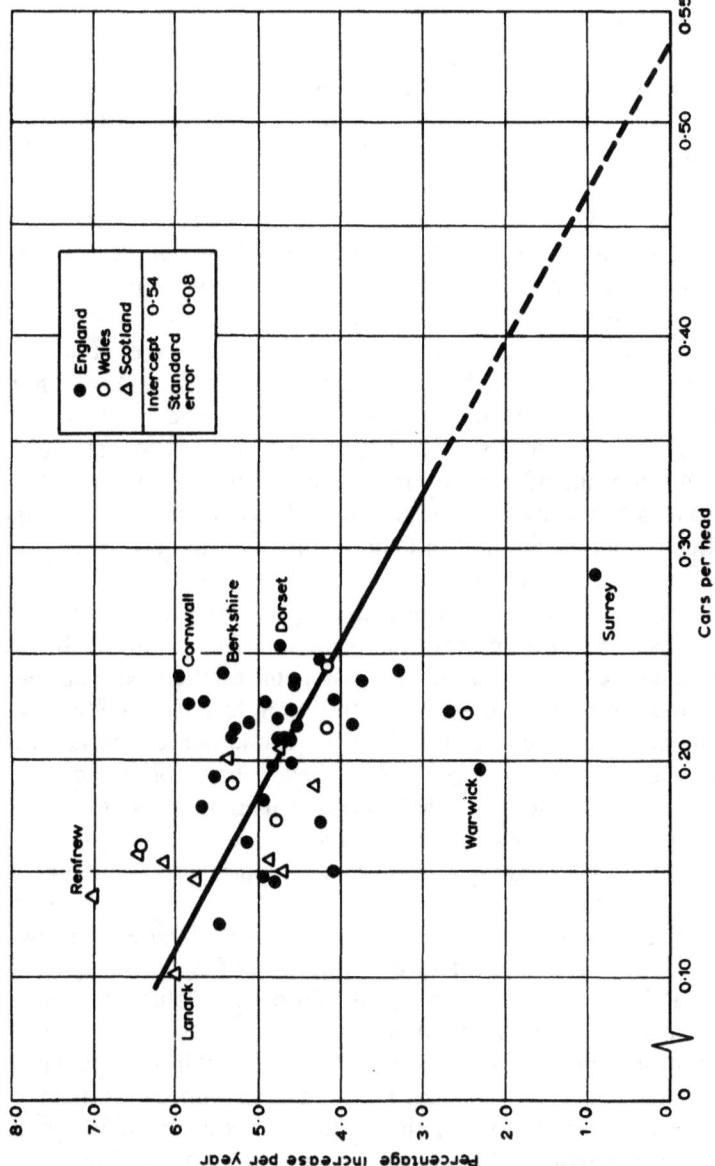

FIGURE 2·1. An example of a graph showing the relationship between growth rates in car ownership and cars per head (for Great Britain 1965–70). (Source: Tulpule, 1973, p. 15.)

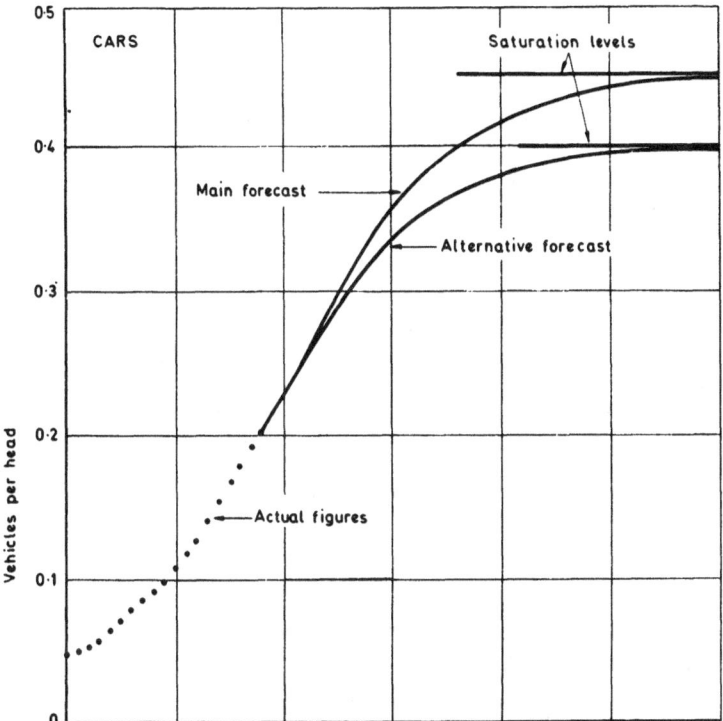

FIGURE 2·2. A typical TRRL logistic curve. (Source: Tulpule, 1969, p. 21.)

been insufficient qualification of the fact that it is most likely a proxy for other socio-economic variables which have demanded more attention. This failure on the part of other workers has led to research of limited scope and value in the field with such obviously inappropriate variables as "distance from London" being introduced as causal factors in explaining regional variations in car ownership.

On the basis of his examination of percentage increase/ownership rate patterns, Tanner was able to postulate a national aggregated saturation level. On the basis of trends in the period 1956–60 he estimated an intercept on the x axis of 0.37 but, taking into account the fact that rates of increase had continued to rise beyond 1960, he decided upon a slightly higher ultimate saturation level of 0.40, with the proviso that as this level had already been reached in several US states and that the latter country was still experiencing considerable growth in numbers. There appeared no reason why the UK should not ultimately experience a far higher saturation level of + 0.5 cars per person.

As in the case of earlier discussion, it is not within the scope of this work to enter into a lengthy debate on all the many social, economic and political factors which could influence the future patterns of car ownership and use. Tanner himself has been noted for his regular insistence that the forecasts should be used with care.

The forecasts are in any case only applicable if there are no catastrophic events such as world wars; if the reader believes that there will be such a catastrophe within ten years from now his own estimates of numbers of vehicles in fifteen years time may be very different from those presented here. It is also assumed in making the forecasts that there will be no substantial substitution by other means of transport (e.g. helicopter, hovercraft) for purposes for which motor vehicles are used at the present time. No explicit mention is made of the use of railways . . . it is further assumed that the lack of road capacity will not always be a major retarding influence on the growth of vehicle numbers. If no further road construction or improvement were to be allowed, then the increases would clearly be much less than if road provision was reasonably closely related to the demand.

Although not mentioned by Tanner, one major limiting feature of the predictions is that no consideration is given to actual population levels and structure, even though this would seriously influence the final number of vehicles in use. Such predictions are left to others. The degree of fluctuation in the field of population prediction has far exceeded any in the field of car ownership per head (see Figure 2·3). Not only does this influence a straight calculation of number of cars on the basis of saturation level — (cars per head × total population) — but the saturation level itself will be influenced by national Gross Domestic Product (GDP) and its distribution amongst the population. In addition, age/sex structure influences economic, physical and legal ability to own or drive a car and so too will household composition. The latter failing has been noted by several economists in part justification for their entry into the car ownership prediction arena.

In his 1962 paper Tanner had suggested that, in the case of cars, there was no evidence to suggest "substantial" variation in ultimate saturation levels between different areas, and that local forecasts of population were required to produce the ultimate rate of cars per head and actual number of vehicles in use. Thus at this stage it was assumed

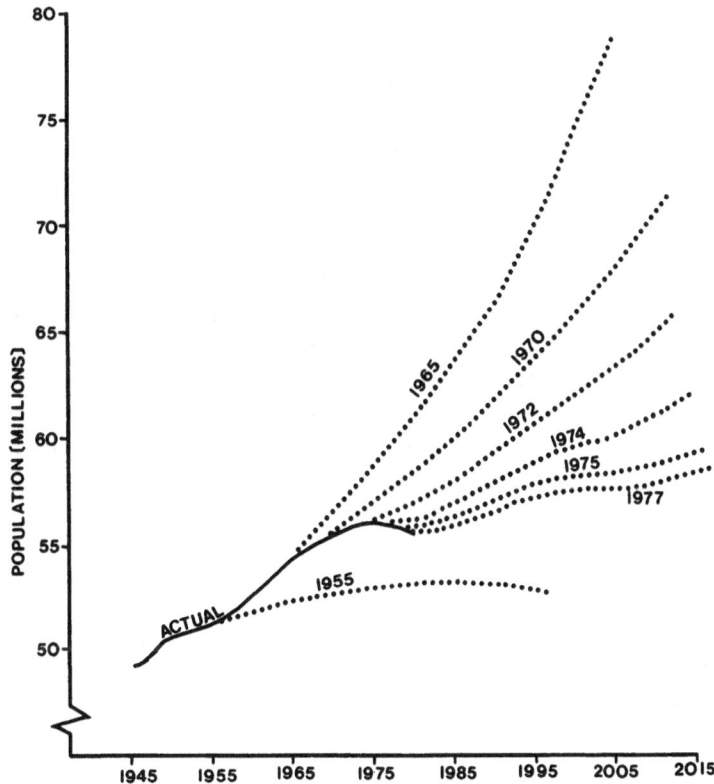

FIGURE 2·3. Alternative past population projections for Great Britain. (Source: Office of Population Censuses and Surveys, *Population Projections*, Series PP2, Nos. 7 and 10.)

that a straightforward application of the national saturation value would provide regional forecasts, disaggregated to the first degree. By 1967, however, Tanner had produced several additional notes on the problem, not only updating predictions at national level on the basis of changes which had taken place 1960–65, but also making provision for local forecasting (at the county level) on the basis of observed variations in ownership and population density. Abandoning an earlier hypothesis (Tanner, 1963) that county levels were significantly related to social class percentages, he suggested a range of saturation level of

0.30 (lowest Greater London) to 0.50 (rural counties). On the basis of this Herrmann (1968) preducted levels of car ownership for all counties, noting that, when aggregated, the county saturation levels came close to a new national Tanner prediction of 0.45. As noted later, both the Tanner and Herrmann figures have been used extensively by local and central government and, while individual predictions have varied, the basic logic behind them has remained unaltered until very recently.

The first notable challenge to the assumptions and methodology inherent in the Tanner approach came as a result of its application as part of the Buchanan Report on *Traffic in Towns* (1963). This resulted in the commencement of what has been a lengthy discussion of the value of economic considerations in the field. *Traffic in Towns* made wide use of the RRL philosophy in predicting levels of car ownership and use within the four sample towns studied. It was in the assumptions and predictions for the largest urban centre (Leeds) that the report and its methodology were strongly criticised. In particular, Beesley and Kain (19674) were harshly critical of the assumption that the general distribution of land use in Leeds would remain static over a fifty year period. They criticised the RRL assumption that the economic influences upon ownership rates could be expected to be limited, thus stimulating intense activity on the part of econometricians into the problem of car ownership. The most powerful criticism centred upon the apparent failure of the report to appreciate the influence of residential density upon car ownership levels. Beesley and Kain were able to demonstrate the significant role which residential density had played in a number of US cities and claimed that a similar situation could be expected for larger British industrial cities. The result was a regression equation relating car ownership to two major variables (residential/population density and median family income). On the basis of this equation, Beesley and Kain produced modified car ownership predictions for Leeds up to the year 2010 (see Figure 2·4), which were notably lower than the Buchanan (RRL) predictions.

In a later paper Beesley and Kain (1965) devoted their entire attention to the problem of car ownership with particular reference to weaknesses which they regarded were characteristic of the RRL method and the way it had been employed as part of the research for *Traffic in Towns*. In particular, the authors were critical of Tanner's vir-

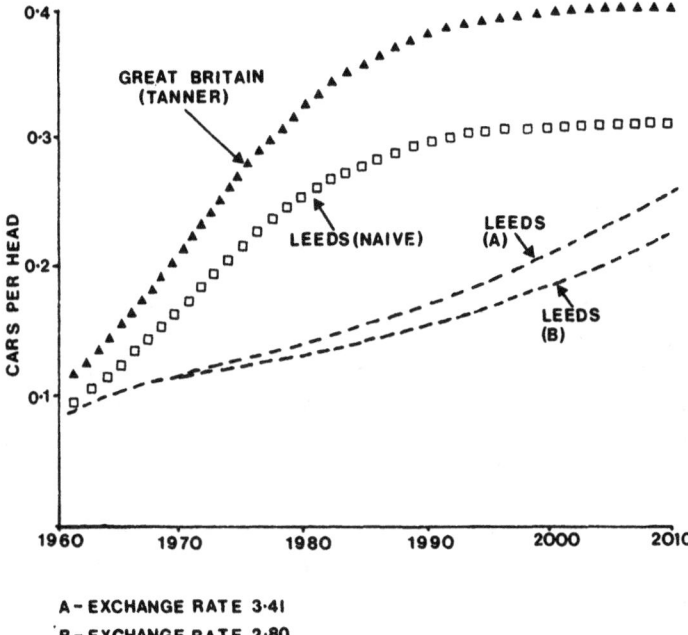

FIGURE 2·4. Comparative car ownership forecasts for Leeds obtained by using different methods. (Source: Beesley and Kain, 1965, p. 167.)

tual omission of economic variables and his "naive" assumptions on patterns of land use and economic development. In 1962 Tanner had commented on the effect of economic limitations being similar to that of congestion, in that both could restrict the number of vehicles but neither would put an absolute limit to the number except an absurdly high one. It was felt that, apart from population density, the only other major factor that might reduce levels in different areas would be economic and social variations. Since social differences were largely caused by economies, and since the forecast levels referred to ownership free from economic limitation, then economic considerations were not considered to be important.

Beesley and Kain felt compelled to "reject" most of Tanner's argument. They claimed that, so long as there were unsatisfied demands or limited incomes, the demand for motor cars, as for other goods, could only be considered in terms of the relative preference for cars, their relative price, the price and availability of complementary goods, and income. There was, consequently, a need to consider future relative prices and "cross-elasticities" between competing goods. Beesley and Kain had tried to do this in their "bundle of goods labelled urban form". Car ownership was seen to be inextricably linked to urban form, and the level of ownership and use had already been shown to be higher in low density residential areas, or in households employed at low density workplaces.

Beesley and Kain included more detailed criticism of the Tanner method for prediction of saturation levels, with the latter's use of very short time range (1955–60) and omission of geographical areas which appeared to invalidate the method. They noted that the increase in the intercept (in three different time periods) in the three UK equations could have resulted from rapid changes in density and distribution of employment since the end of the war. In a similar fashion the larger intercepts evident in US regressions could be explained in terms of different urban form in US cities compared to their UK counterparts (see Figure 2·5).

It was claimed that the saturation level was even more dependent upon density and distribution of employment and population, and on quality of public transport. Consequently, when these additional factors were considered, "the idea of a saturation level loses its grand simplicity".

Does this mean that Tanner's forecasting method and forecasts are invalid? It does suggest they should be used with considerable care, and especially for periods beyond 15–20 years. His results *might* be correct if the estimated saturation level reflects the on-going trend adjustments in urban form and other determinants of car ownership, and if these trend adjustments continued at historical rates. However, the evidence of substantial and rapid change in the UK intercept over just 14 years obtained by the method — from 0.16 *per capita* for the period 1946–51, to 0.22 *per capita* for the period 1956–60, weighs against this. (p. 185)

Following the criticism which followed the publication of *Traffic in Towns*, the RRL extrapolatory method was modified to allow for variation in residential density, and the changes were evident in Herrmann's county predictions (1968). This represented the first of a number of adaptations made by Tanner to his basic original model,

THE ROAD RESEARCH LABORATORY MODEL 27

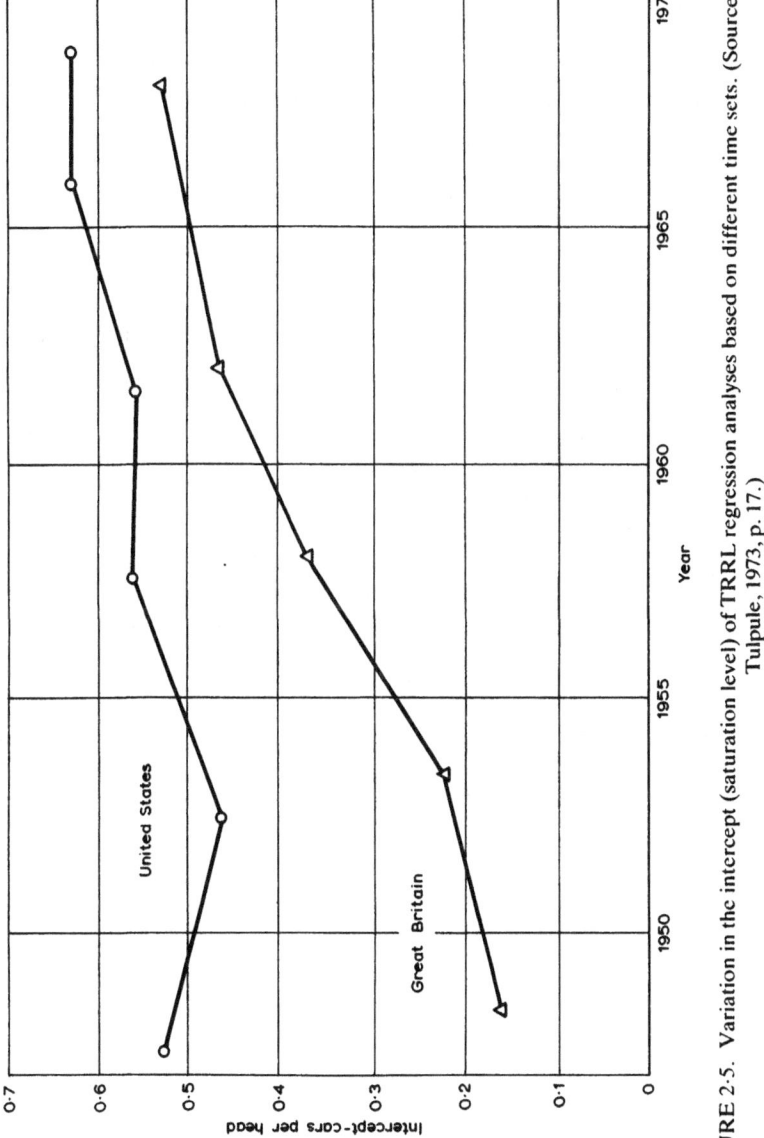

FIGURE 2.5. Variation in the intercept (saturation level) of TRRL regression analyses based on different time sets. (Source: Tulpule, 1973, p. 17.)

but the idea of a graphical extrapolation of past and present trends has remained the central core of the approach and as such it has been consistently accepted by central and local government as the official basis for predictions. More recent criticism of the inherent assumptions of the RRL methodology are contained in Chapter 4. In the meantime, it is worthwhile examining some of the central themes behind the econometric approach to car ownership prediction which have received widespread use in transportation studies.

REFERENCES

Beesley, M.E., and Kain, J.F., Urban form, car ownership and public policy: an appraisal of 'Traffic in Towns', *Urban Studies*, I, No. 2, 1964, pp. 174–203.

Beesley, M.E., and Kain, J.F., Forecasting car ownership and use, *Urban Studies*, II, No. 2, 1965, pp. 163–185.

Herrmann, P.G., Forecasts of vehicle ownership in the counties and county boroughs of Great Britain, *Road Research Laboratory Report LR200*, Road Research Laboratory, Crowthorne, 1968.

Tanner, J.C., Forecasts of future numbers of vehicles in Great Britain, *Roads and Road Construction*, XL, 1962, pp. 263–274.

Tanner, J.C., Car and motorcycle ownership in the counties of Great Britain in 1960, *Journal of the Royal Statistical Society* (A), 1963, pp. 276–284.

Traffic in Towns: a study of the long term problems of traffic in urban areas, HMSO, London, 1963.

Tulpule, A.H., Forecasts of vehicles and traffic in Great Britain, *Road Research Laboratory Report LR 288*, Road Research Laboratory, Crowthorne, 1969.

Tulpule, A.H., Forecasts of vehicles and traffic in Great Britain, 1972 revision, *Transport and Road Research Laboratory Report LR 543*, Transport and Road Research Laboratory, Crowthorne, 1973.

3 Econometric Modelling and the Problem of Disaggregation

ECONOMETRIC MODELLING, so called because of the basic discipline of early protagonists, aims to develop and calibrate mathematical models which explain a particular aspect of consumer behaviour. In this case the problem is concerned with explaining, in terms of causal relationships, why people buy cars. Often the analysis aims to extend the explanation to different categories of purchaser and vehicle. The statistical parameters of these models are usually estimated and tested using data in time series or cross-sectional form. Thus, some models have examined the relationship between car ownership and a number of independent variables over time; others have used data related to a specific area at a specific time, with the intention of using the derived model as a basis for future projections, based on independent estimations of change in causal variables.

Causal modelling is primarily based on the concept of utility theory. The latter, at household level, attempts to place some quantitative assessment on the value placed on ownership of goods or services. This often involves some numerical estimation of the value to individuals of certain intangible phenomena such as quality of life and level of accessibility, relating these to some form of generalised cost compared to the individual's ability or need to afford them, usually in money terms. The theory formed the basis of the earlier work of Mogridge (1967 onwards) in which he sought to introduce econometric reasoning into the extrapolatory modelling procedure. By doing so he offered a challenging alternative saturation level and long-term forecast of car ownership to that which had been proposed by Tanner, albeit with an argument which contained several basic weak assumptions. His essential argument (1967) was that the average individual of a given money income level will have a pattern of expenditure on

30 CAR OWNERSHIP FORECASTING

various categories such as food, housing, and fuel, and that this pattern is stable over time for a given society. Patterns of change in spending can be explained by changes in income distribution. Although admitting that expenditure on housing and fuel were liable to fluctuation, Mogridge grouped transport with food as examples of static expenditure items. With the benefit of post-1973 hindsight, it is possible to criticise the logic of this assumption. The repeated basis for the reasoning was that expenditure patterns in each commodity group are related to income levels. Once income levels, distribution, and growth rates are established, it is possible to hypothesise on patterns of total expenditure likely to be experienced in each commodity group. So it was assumed "that the average person of a given income spends a given quantity of money on the purchase of cars and that there is a function

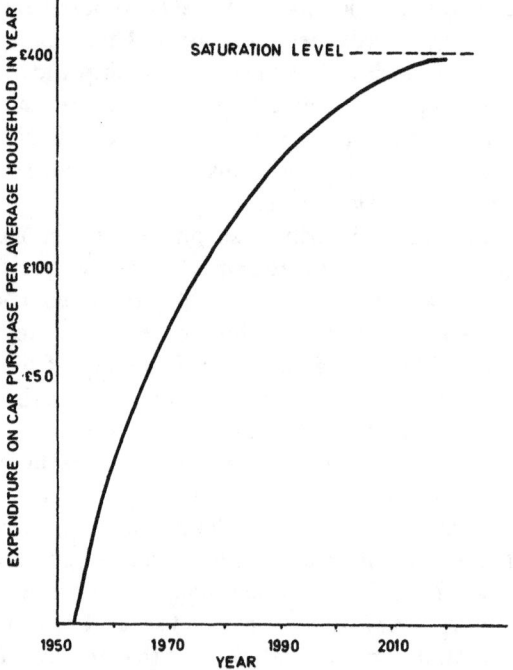

FIGURE 3·1. A forecast of expenditure on new cars 1967–2010. (Source: Mogridge, 1967, p. 72.)

which gives, for every income level, an average expenditure on car purchase."

Mogridge, like Tanner before him, went on to surround his work with a number of assumptions, which, although apparently valid in 1967, have since been proved to have been optimistic or unfounded (such as continued UK growth rate of 6 per cent in the number of business-owned cars etc.). On the basis of past trends, he was able to produce a cumulative log-normal curve which eventually reached a saturation level of 0.66 cars per person, much higher than Tanner's (current) 0.45. The Mogridge level was based on the expectation that, since income influenced ownership and income levels were changing, by the year 2000 most individuals would be in a much higher income category with resultant higher rates of car ownership. Despite a similar trend up to 1980, from then onwards the Mogridge curve continued to increase at a constant rate before reducing and eventually terminating at the higher saturation level (see Figures 3·1, 3·2).

Mogridge's methods, and conclusions, were vigorously criticised by Evans (1970), particularly with regard to the assumption that the income/car purchase relationship was static. Evans rejected the idea that an increase in the average real income of households at a given initial income level (bringing them to a higher income level) will lead them to adopt the expenditure patterns of those households initially at the higher income level. He believed that households at different points on the income scale would exhibit differences in composition and characteristics and that they would not share equal priorities on the manner in which they allocated their consumer spending. Evans also suggested that Mogridge's over-emphasis on income as a causal factor in ownership had produced an over prediction of saturation level.

Category Analysis

In the meantime, several parties had become involved with the application of category analysis techniques to the problem of car ownership prediction. Not only was it felt that extrapolatory procedures contained too many imponderables and unrealistic assumptions, but the category analysis technique provided the degree of disaggregation necessary to make the results applicable at the local and household

32 CAR OWNERSHIP FORECASTING

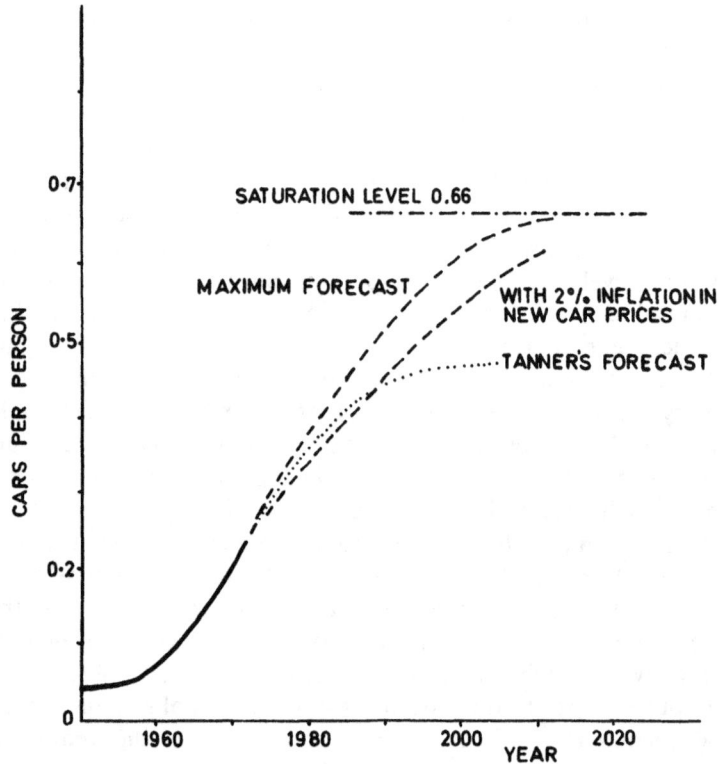

FIGURE 3·2. Alternative forecasts of car ownership by Mogridge. (Source: Mogridge, 1967, p. 72.)

level where they were of most benefit in regional transport planning studies. The work of Wootton and Pick (1967) provided an early and conventional example of the application of the technique. The latter involved the use of a multi-dimensional matrix, each dimension representing an independent variable, the values of which are grouped into a number of classes or categories. Categories are not necessarily numerical but all form discrete groupings into which all data must be inserted. The classes are used to cross-classify households and the average (aggregated) level of car ownership for households exhibiting a certain set of characteristics is obtained. It is then a fairly straightforward procedure to estimate the number of households likely to be present in each cell (at some future "design" date) and project their likely car ownership levels.

Wootton and Pick were primarily concerned with a procedure for the estimation of trip generation but saw prediction of car ownership levels as a major factor in this stage of the transport planning process. On the basis of data from the West Midlands they were able to construct conditional probability distributions of car ownership against income (probability of 0, 1, + 1 cars per household for a given income). Figure 3·3 shows a typical example in graphical form. The West Midlands findings were tested against comparable cities/conurbations in other parts of the world and on the basis of an acceptable fit between expected and observed levels of car ownership for the period 1953–65 a forecast of cars per household up to 2010 was produced. It was assumed that, for projection purposes, the functions $P(n/x)$ would be stable over time and that the income distribution relative to the price of cars would change.

In addition to hypothesising upon the relationship between income and car ownership, Wootton and Pick also gave limited consideration to the influence of population density and public transport provision (accessibility indices), although the latter was noted more in terms of its influence on trip generation than on levels of car ownership.

It should be noted that Wootton and Pick produced their predictions in terms of cars per household. As in earlier attempts at car ownership prediction, they left to others the all important matter of population/household size/structure prediction. Thus, in real terms, this attempt at disaggregation was of limited practical value. Although the problem of population prediction is beyond the scope of this work, that does not alter the fact that the problem should be more carefully considered by those working in the travel prediction field. With regard to this matter, Wootton and Pick concluded on a misguidedly optimistic note when they stated that, on the basis of observed similarities in travel habits between superficially dissimilar areas, the method could lead to a rationalisation of survey methods, and that, with a few sample surveys, coupled with census data, estimates of travel patterns would be an "immediate reality".

The category analysis concept was applied widely in a number of transport studies in the late 1960's. A computer package produced and

34 CAR OWNERSHIP FORECASTING

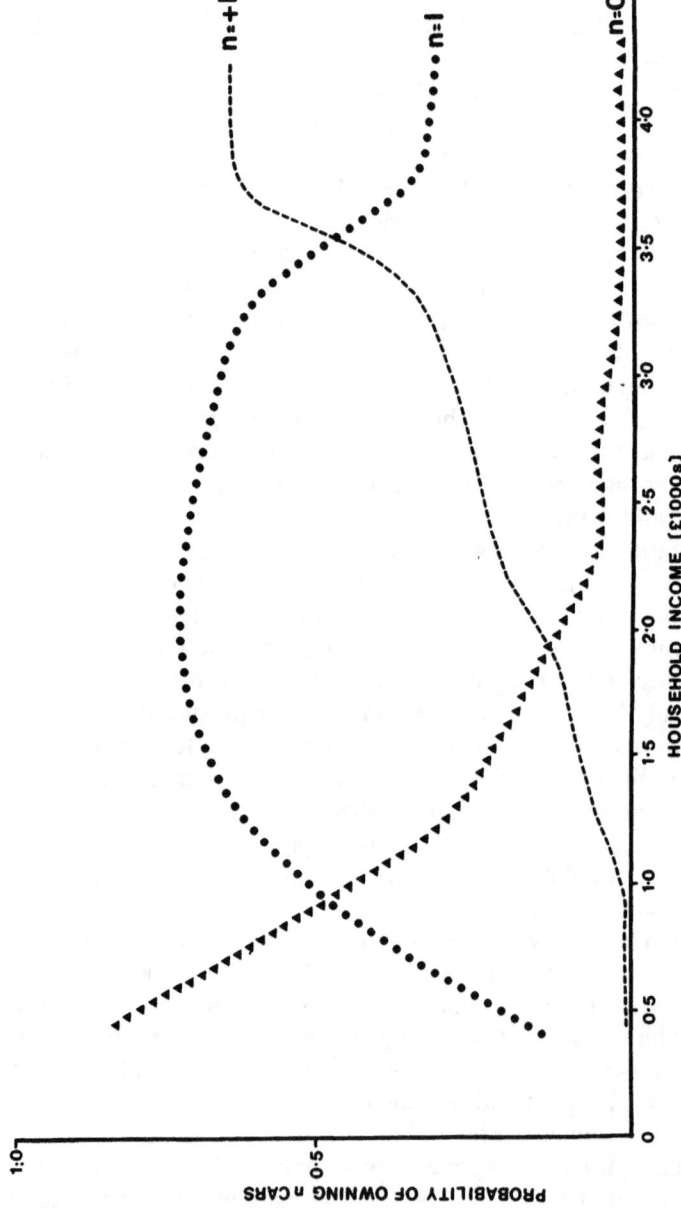

FIGURE 3·3. A typical set of car ownership/income probability curves.

refined by Transport Planning Associates (TPA) has been used extensively at regional level in the UK. This contains a car ownership submodel similar in concept to the Wootton and Pick procedure.

Mogridge (1970) (with Eldridge) developed his earlier work into a model which was used by the Greater London Council (GLC), incorporating the variables of income, density and employed residents per household as determinants of car ownership. This involved the introduction of a new concept of income, namely car purchasing income, but the basic hypothesis, that the income/car ownership relationship remained stable over time, remained. In addition, Mogridge acknowledged the need to consider a saturation level for household car ownership since, taken to the ultimate conclusion (at the levels of economic growth experienced in the 1960's), ownership levels could be projected far in excess of those already extant in the more affluent regions of the USA.

Much of the modified philosophy of Mogridge and Eldridge has remained in official GLC policy. Mogridge noted that Tanner's national saturation level of 0.45 (latest estimate in 1970) was too low and, on the basis of the probable future number of driving licence holders in the target year (95 per cent men and 80 per cent women able to drive), produced a saturation level of 0.60 cars per qualified driver. This was reduced to 0.55 after taking account the effect of children and the aged. It was assumed that no person would purchase more than one car and taking into account the effect of limited parking facilities and congestion, an upper limit was placed on household car ownership of 1.6.

It is not very difficult to criticise some of these assumptions and several authorities have done so at length. Fowkes and Button (1977) have criticised the assumptions behind the (arbitrary) hypothesised saturation level. In particular, they have challenged the (aggregated) 1.6 cars per household limit and the assumptions that only 80 per cent of women but 95 per cent of men would be "capable" of driving and, given this, the assumption that all (potential) capable drivers would become qualified and purchase one car each.

In addition, the introduction of the new concept of "car purchasing income" has caused lengthy comment and examination. The calculation of values for this new variable involved the addition of the (expected) percentage fall in car prices to the conventional forecast of income growth. During the period immediately preceding the development of the Mogridge/Eldridge model, new car prices were "falling" in real terms by 1.3 per cent per year, whereas incomes were growing at 2.3 per cent per year, thus resulting in a "car purchasing income" increase of 3.6 per cent per year.

The most notable critic of this use of car purchasing income was Bates (1971). His own interpretation of car price trends during the late 1960's led him to conclude that such prices had in fact increased and that Mogridge had placed too great a weighting on this variable as an explanation of the increase in car ownership. Bates provided independent estimates of income and car price changes for the period 1966–70. Whereas personal disposable incomes had increased at 4.84 per cent per annum, car prices had increased by 5.19 per cent during a period when the consumer price index had only risen, each year, by 4.11 per cent.

Following a "hard look at car ownership modelling", Bates was of the opinion that one of the fundamental asumptions of most previous econometric/category analysis work, that the income/car ownership function is stationary over time, was in fact invalid and that, as a result, these expensive models were failing to produce reliable forecasts. In addition it was argued that the practice of sampling income from a few zones and then using observed car ownership rates to produce general estimates of income was an admission that the model was not working correctly.

Bates also noted the problem of regional variations in car ownership levels and growth rates. (It is surprising that at this stage little attention was being given to the necessary problem of disaggregation of national forecasts and models to make them applicable and reliable at regional and local level.) In response to notable variations in regional ownership levels, Bates noted that existing econometric models could not provide a satisfactory explanation and was forced to conclude that the most satisfactory method would be a RRL type national extrapolation modified at regional level by consideration of population density variations. Since much of Bates' work was based on national (English) data disaggregated only to the first degree of eight planning regions, the problems resulting from predictions for each of these regions (each with its own large variation in population density and distribution) were sufficient to invalidate any worthwhile results at that stage.

The result of Bates' own work in the early 1970's was the production of a new model using an extrapolatory "quasi-logistic" curve (using data from 1965 National Travel Survey) which took the form:

$$\log \frac{P_0}{1-P_0} = a + b \log Y \qquad (3.1)$$

where Y is income at current prices and P_0 is the probability of a household not owning a car.

The above is simplified to:

$$P_0 = \frac{cY^b}{1+cY^b} \quad \text{where } c = e^a \qquad (3.1.2)$$

This new model has been extensively tested by Bates and others and subsequently extended to include the effects of residential density, the importance of which had been noted earlier. It was concluded that an explanation of ownership rates for single (P_1) and multiple (P_2) car owning households is found from

$$\frac{P_2}{P_1} = a_1 e^{-b_1 Y D^{c_1}} \qquad (3.2)$$

$$P_0 = 1 - (P_1 + P_2) \qquad (3.3)$$

where D is the number of persons per acre.

This attempt to introduce residential density as a control on prediction has been criticised by Fowkes and Button (1977). They argue that density (as a proxy for other factors such as public transport availability, congestion levels, parking restraint) may have some effect on national saturation levels but is not necessarily an appropriate variable for incorporation into models applied in smaller discrete areas, because it is "imprecise". In addition, it is claimed that residential density is closely correlated with income and, to an extent, household structure.

This criticism is valid only to a degree. It would appear that equal skill is required in a sensible reasoned choice of study area as in a choice of variables where some degree of local homogeneity exists. Unfortunately, many administrative areas are deliberately shaped to provide a social and economic (and density) mix for purposes of finan-

38   CAR OWNERSHIP FORECASTING

cial equity. This, together with a gradual breaking down of the existing correlation between density and income, requires research to progress a stage further and examine the variables for which density has been used as a proxy.

This has already been attempted by a number of workers. Fairhurst (1975) took, arguably, a much more pragmatic approach when he examined the crucial role which public transport provision appeared to play in influencing car ownership in London. (It should be noted that London provides an "extreme" example of the manner in which such socio-economic variables influence car purchasing patterns and car use and a straightforward linear extrapolation of trends to smaller cities — especially non-conurbations — is not valid.) Fairhurst used the concept of a public transport acecssibility index based on acreage of zone/mid-day frequency of public transport services, but was careful to outline all the multifarious factors influencing car ownership and the complexity of the task. (Some of these are evident in Figure 1·1.) The result was a modification of the earlier Bates model and a conclusion that, since public transport provision had been shown to be a statistically significant factor in car ownership modelling, changes in scale and pattern of provision of the type already initiated by central and local government were an important influence on projections. The converse effect of car ownership levels on public transport, which lies beyond the scope of this book, has been the subject of much recent research by TRRL staff and others. This has been primarily aimed at employing causal relationships to estimate future levels of public transport use with particular reference to evidence of a "bottoming out" of the decline in demand in mirror image to saturation in car ownership.

Regional Variation

As work has continued, an increasingly apparent problem has been the spatial disaggregation of models and projections to produce practical guidelines at the local level. Although recent developments are outlined later, is appropriate that the work of a number of commentators in relation to the English regions is noted here.

The problem of appreciation of regional variation in ownership

ECONOMETRIC MODELLING 39

FIGURE 3·4. Regional averages for cars per thousand, average incomes and population per square mile, 1966.

Region	Averages			Rank order (descending)		
	Cars per thousand of population	Income per case assessed £	Persons per square mile	Cars per thousand of population	Income per case assessed £	Persons per square mile
South Western	213	1,033	406	1	4	10
East Anglia	212	1,028	326	2	6	11
Wales II	209	971	153	3	14	12
South Eastern	202	1,164	1,608	4	1	2
Southern Scotland	191	1,000	58	5	10	13
Northern Scotland	185	975	45	6	13	14
East Midland	183	1,053	688	7	3	7
West Midland	172	1,089	999	8	2	3
Yorkshire and Humberside	163	1,013	864	9	8	5
Wales I	162	995	643	10	11	8
North Western	151	1,033	2,302	11	4	1
East Central Scotland	142	1,016	836	12	7	6
Northern England	139	989	450	13	12	9
West Central Scotland	114	1,002	998	14	9	4

(Source: Sleeman, 1969, p. 313.)

rates (and the factors causing this variation) had been noted by a number of observers during the 1960's. Sleeman described variations in geographical distribution of motor cars as early as 1961 and engaged in a "new look" at the same problem in 1969 which included a new survey of variations at the local authority level. Figure 3·4 summarises the degree of variation at planning region level. Sleeman sought to explain regional variation in terms of population density and average income and established two regression equations, one for "less urbanised areas" and the other for "more urbanised areas". Although outlining the need for study on a local basis, Sleeman's work, by definition, did not extend far enough down the hierarchical chain of spatial disaggregation, and he was content merely to provide a raw list of ownership rates at the county/county borough level, the degree of variation within which only exposed the limitations of his earlier regional work.

Following on from Sleeman, Buxton and Rhys extended the scope of the examination, introducing age structure of the local population level as an additional independent variable and placing more scientifically based numerical limits on the terms more urbanised/less urbanised. While supporting the general methodology and conclusions of his work, they found difficulty in explaining the strong correlation between ownership and latitude (distance north, which Sleeman hypothesised was a proxy for climate). They justifiably criticised any possibility of future transport policy being influenced by an assumption as tenuous as this. A more satisfactory explanation seems to be basic geological structure and the increasing occurrence of formerly economically-important minerals (northwards), the unplanned exploitation of which has resulted in the type of contemporary, socio-economic characteristics which were in fact the real cause of the variations in car ownership. This involves consideration of economic theory and the "fringe area" concept which is beyond the scope of this discussion, but which does expose the need to employ objective reasoning in the problem of examination of statistical relationships.

Button appeared particularly aware of the problem of use of aggregated data for predictions at local level (a problem which has received increasing attention since) stating that the problem had:

> two aspects; analysis and forecasting. Excessive areal averaging of variables reduces be-

tween-zone variation at the expense of increasing the variance within each zone with the result that regressions based upon these zonal averages tend to yield an over-optimistic impression of the actual variation in household behaviour being accounted for. A supplementary point is that lack of uniformity within zones may result in an error term which is neither normally distributed nor has a constant variance and consequently violates a basic assumption of least-squares regression — that of homoscedasticity.

The second problem is that parameters derived from models employing aggregated data can only be assumed to remain constant through time if the internal structure of the zones is unchanging. As such an assumption is unjustified, for all but short time horizons, models of this type have limited forecasting potential. (Button, 1973, p. 76).

It is clear that Button was touching upon a hitherto neglected cause for concern in previously adopted prediction techniques, namely the failure to allow for possible temporal changes in socio-economic conditions in spatially discrete areas which had the type of unchanged geographical boundaries which might suggest social and economic "stability" on the part of the population.

Working on data at local authority area level in the West Riding of Yorkshire, Button carried out conventional multiple regression analysis, employing the widely accepted independent variables of household size/employed population/social characteristics/rate of home ownership/population density as an explanation of level of car ownership. Local authority data produced a multiple coefficient of determination (R^2) of 80.28 per cent. When separate analyses were carried out using the same variables at the ward and parish level the R^2 values fell to 47.34 per cent and 33.4 per cent respectively drawing the conclusion that:

despite the lower coefficients of multiple determination the reduction in grouping suggests that the disaggregated models provide a more realistic picture of the underlying causes of divergences in car ownership rates. (p. 77).

The theme of geographical variation in ownership rates (and, therefore, in variables influencing these rates) was also taken up by Fishwick (1972). He noted a number of independent variables which were expected to influence car ownership and exhibit variation at the regional and local level. These are:

1) Population density
2) Total rateable value per head of population

3) Total personal income per head of population
4) Mileage from London
5) Percentage of economically occupied males in selected social groups.

An examination at county level of the relationship between these variables and car ownership using multiple linear regression produced a number of notable results though their significance was tempered by the fact that most independent variables were also closely correlated with other significant variables as well as the dependent variable. It is clear that part of the reason for this is the degree to which variables such as population density and distance from London are in fact proxies for other variables, which may include total rateable value, personal income and percentage of economically occupied males. Thus, there is a certain degree of duplication of real independent variable which has had the effect of reducing the validity of the exercise.

REFERENCES

Bates, J.J., A hard look at car ownership modelling, *Mathematical Advisory Unit Note 216*, Department of the Environment, London, 1971.

Button, K.J., Motor car ownership in the West Riding of Yorkshire: some findings, *Traffic Engineering & Control*, 15, June, 1973, pp. 76–78.

Buxton, M.J., and Rhys, D.G., The demand for car ownership: a note, *Scottish Journal of Political Economy*, June 1972, pp. 175–181.

Evans, A.W., The prediction of car ownership — a comment, *Journal of Transport Economics and Policy*, IV, 1970, pp. 89–99.

Fairhurst, M.H., The influence of public transport on car ownership in London, *Journal of Transport Economics and Policy*, IX, 1975, pp. 193–208.

Fishwick, F., The influence of economic factors on car ownership in Great Britain, *Proceedings of PTRC Seminar on Urban Traffic Model Research*, PTRC, 1972.

Fowkes, A.S., and Button, K.J., A survey of techniques of car ownership forecasting, Institute for Transport Studies, University of Leeds, Working Paper No. 92, 1977.

Mogridge, M.J.H., The prediction of car ownership, *Journal of Transport Economics and Policy*, I, 1967, pp. 52–74.

Mogridge, M.J.H., and Eldridge, D.A., Car ownership in London, *Greater London Council Intelligence Unit Report No. 10*, 1970.

Sleeman, J.F., The geographical distribution of motor cars in Britain, *Scottish Journal of Political Economy*, February, 1961, pp. 71–81.

Sleeman, J.F., A new look at the distribution of private cars in Britain, *Scottish Journal of Political Economy*, September, 1969, pp. 306–318.

Wootton, H.J., and Pick, G.W., A model for trips generated by households, *Journal of Transport Economics and Policy*, I, 1967, pp. 137–153.

4 The Impact of Government Intervention

THE PERIOD during which the described developments in econometric modelling were taking place was one of increasing government involvement in transport planning. Not only was there a steady stream of government directives on car ownership prediction, but most conurbations and freestanding towns were being subjected to intensive statutory transportation surveys. From these were produced elaborate projections of transport demand and the necessary infrastructural and administrative changes which were considered necessary to accommodate the expected patterns of movement. A number of these studies were using econometric modelling (particularly category analysis) at the very time that academic debate over its validity was most active.

At the national level, the work of Tanner had gained government acceptance and formed the basis of Ministry of Transport (later Department of the Environment) memoranda for use by highway planners. A major development in highway planning was the introduction of the COBA cost-benefit computer analysis package, which, by 1974, local authorities were under statutory directive to use in conjunction with highway schemes with capital cost in excess of £500,000. Part of the COBA input took the form of car ownership predictions using virtually the same logistic curve/saturation level concept as that propounded by Tanner in 1962 and altered, to take into account variations in residential density, in 1967. Thus, whilst the debate on econometric analysis was continuing, the RRL forecasting techniques remained largely unchallenged (at least in terms of methodology, rather than basic philosophy) and were the subject of widespread application at national and local level as partial justification for an intensive schedule of major highway projects. This was in spite of Tanner's earlier reservations concerning the necessary economic and social conditions/

trends required to ensure the actual materialisation of his projected pattern of car ownership. There is little doubt that interested parties were particularly impressed by the way in which, as the years passed by, the Tanner predictions at national level provided a remarkably close fit with actual levels of car ownership. This tended to disguise the fact that the main failure of the use of the extrapolatory logistic curve, in the eyes of many commentators, was the method by which Tanner had calculated his saturation level and his assumption that such a state of affairs (0.45 cars per person) was a practical proposition in this densely populated country. With the growth of criticism of the use of mathematical modelling in transport planning in general, and the use of aggregated extrapolatory prediction procedures in particular, the stage was set for a renewed discussion on not only the detailed mathematical strengths and weaknesses of the TRRL method, but the wider issue of its use and abuse.

The oil crisis of late 1973 occurred while the TRRL team were concerning themselves with another revision of the national car ownership predictions. Wage and price inflation was already having some impact upon the pattern of relatively stable growth in economic and social activity upon which several earlier assumptions had been based. In addition, it was clear that earlier predictions on population growth and future household structure/income were, on the basis of more recent trends, very wide of the mark. Accordingly, there is evidence of a belated acknowledgement of the changing energy supply and general economic situation in the TRRL 1974 revision of forecasts of vehicles and traffic in Great Britain with the inclusion of a "high", "middle" and "low" forecast (see Figure 4·1). Each forecast was based on certain assumptions about growth of GDP and motoring costs: thus 1974 marked in a limited fashion a break away from the earlier rigid pattern of a single curve and single saturation level at a single design year. It is clear that Tanner was still deeply influenced by the example of the USA. Using his "conventional" method for calculating saturation level by regressing rate of change in car ownership growth on actual ownership rates for each state he had noted a remarkable stability in the value of the intercept (saturation level) when applied to different time periods (see Figure 4·2).

Nevertheless, in taking into account a greater number of factors than in earlier studies (notably the effect of slower economic growth

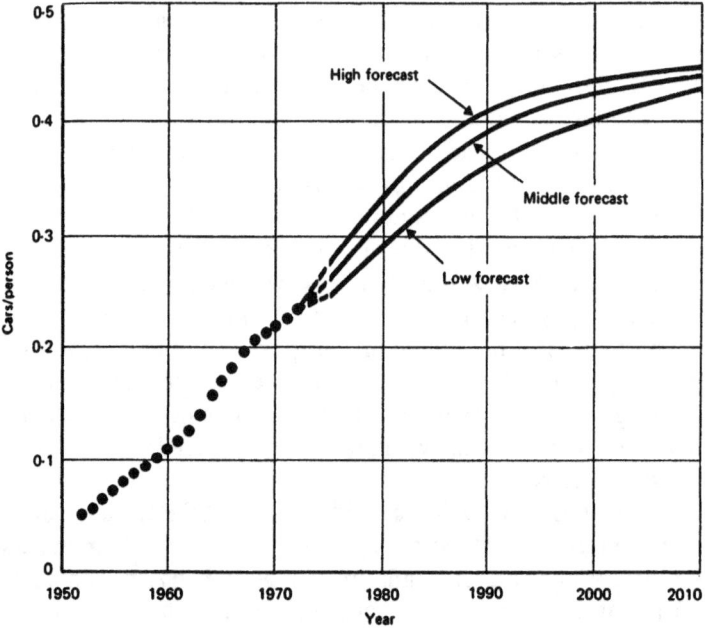

FIGURE 4·1. Cars per person in Great Britain: actual 1952–1973 and forecast 1975–2010. (Source: Tanner, 1974, Figure 1.)

rate, different age distribution in the population and new assumptions in the nature of and influence of population density), Tanner decided to retain his earlier saturation level noting that none of the methods used:

> gives by itself a convincing estimate of saturation level, and the estimates vary somewhat. Of the calculations based on figures for Great Britain, one (national time series) gives figures of about 0.3 to 0.4, while two others (intercept method and age distributions) suggest higher values, 0.5 or more. All three calculations based on United States data suggest a fairly high saturation level at least for the United States. The international comparison gave no clear estimate.

It is also worthy of note that, although there had been some re-examination of the method used for calculation of saturation levels, the use of the logistic growth curve remained virtually unchanged from its first application in the early 1960's.

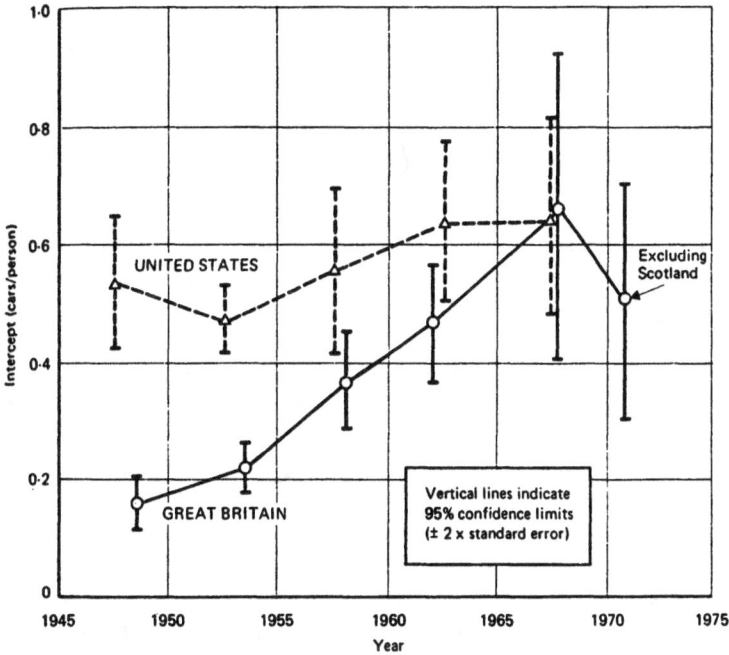

FIGURE 4·2. Variation of intercept for cars per person, Great Britain and United States. (Source: Tanner, 1974, Figure A1.8.)

The results of the 1974 LR 650 report were duly accepted by the Department of the Environment and, taking into account the earlier county estimates of Herrmann, (and the methods used for their calculation), they were issued to local authorities and central government organisations in the form of DOE Technical Memorandum H3/75. In accordance with the acceptance of an unchanged saturation level, the basic input into COBA remained unaltered.

Coming as they did at a time of great uncertainty over future supplies and price of fuel, population growth and economic development, and marked by a tendency to rely on existing concepts and methods based on experience in the 1950's and 60's, it is not surprising that the 1974 forecasts triggered renewed debate on the role of extrapolatory modelling in the field of transport planning. Most of the criticism has been concerned with the method for estimating saturation levels. There are assumptions of linearity in the relationship between the two variables used which are, because of the use of extrapolation, difficult to prove, or at least justify. The use of US states or UK counties as data bases introduces a spatial source of variance into the study. The population of each statistical area cannot be assumed to be homogeneous in terms of socio-economic characteristics or spatial distribution (density). Taking this a stage further, it cannot be assumed that if a correlative relationship exists (a) between rate of growth and (b) cars per person, then this takes place in a vacuum totally independent of other socio-economic factors. Indeed, it is other socio-economic factors which are causing the variation, yet they are not identified scientifically. Tanner's own examination of the effects of the 1930's depression and the Second World War showed the possible extreme influence of exogenous variables upon which both the rate of growth and the car ownership rate are dependent.

The most vociferous criticism of the Tanner methodology of late has been provided by the geographer Adams (1974). He laid particular emphasis upon the method used to estimate the third parameter of the logistic curve (saturation level) and stated that its validity depended upon the "bold" assumption that a relationship derived from the spatial variation in growth and ownership could be interpreted as a temporal trend which would remain constant over a period of forty years.

Adams (in fact pre-dating the 1974 revision of Tanner but criticising procedures which were being retained in the TRRL work) went on to suggest that, in the regression analysis used by Tanner, car ownership should be the dependent (y) variable whereas rate of change should be the independent (x) variable and that if y was then regressed on x a saturation level of 0.28 cars per head would be given . . . "just over half the level predicted by the TRRL regression of Tulpule" (who had used the Tanner method in his 1972 forecasts).

Adams also criticised the failure of both Tanner and Tulpule to provide correlation coefficients as an indication of the strength of the relationship between the two variables used. Tanner (1974) replied that, as the error structure inherent in growth rates was far greater than that existent in the measurement of actual ownership rates, it was in fact a valid exercise to use ownership as the "independent" and growth rates as the "dependent" variable. However, as noted earlier, the entire exercise could hardly be described as a piece of "conventional" statistical methodology but merely a practical means of calculating the essential third parameter of the logistic curve.

Although Adams' criticism regarding the arrangement of variables has been effectively countered by Tanner (and others), his conclusion, that each separate county is likely to have its own saturation level, is valid, because each is still likely to have variations in population density if not other socio-economic factors, and such variations have been admitted even by the most ardent supporters of the time-series extrapolatory approach. The attempts at disaggregation noted earlier have been a part response to this.

Adams (1975) also criticised the over-employment of "short range" cross-sectional regression analysis in the TRRL procedure and, despite Tanner's claims that the cross-section regression method had given consistent results, a summary diagram showing the "history" of the results of such methods (which had used data 1946–72) exposed the wide variation in results which had been produced. Such procedures had tended to extrapolate on the basis of trends evident in periods as short as two years and were therefore easily exposed to error. Over the longer term, however, (1952–1973) there is strong evidence of a distinct negative linear trend in the relationship between percentage increases and ownership for given years. It is fair to note that Adams did not offer any constructive alternative solution to the problem of numerical prediction of car ownership, having initially entered into the discussion on the grounds of basic philosophical disapproval of the general use of mathematical modelling in transport planning.

Fowkes and Button, who have already been noted in the context of their support for econometric as opposed to time series procedures, have noted that, in any case, the use of the cumulative logistic curve is open to question and they may have known that Tanner was already working on alternative curves when they provided their reservations (in 1977). Noting that the logistic curve proved useful in explaining growth rates for television sets they concluded that motor cars could not be expected to have a similar rapid rate of market absorption. Despite the fact that Bonus (1973) had shown that car ownership in post-war West Germany had followed a logistic growth pattern, it could be concluded that this was due to economic factors rather than market diffusion. On several occasions Mogridge has criticised the analogy between biological diffusion — upon which the logistic curve is based — and that experienced in the area of car ownership. Fowkes and Button concluded that econometrics can provide a valid *explanation* of car ownership growth in addition to being a useful descriptive tool. Not surprisingly, most of their work has been in support of this philosophy, and the application of what could be termed causal models.

Following on from the work of Adams there was further discussion of the relative validity of TRRL forecasting methods, with particular reference to the inappropriateness of the method used for prediction at regional level. Kirby (1976a), with the benefit of having studied Tanner's 1974 report, introduced the work of Ioannou (1975) into the discussion concerned with the overall validity of the techniques employed. The result of his own analysis produced some significant observations on the problem, with particular reference to the characteristics of the data sets used to estimate saturation, the method of regression employed, and the overall appropriateness of the application of a single rigid saturation level and logistic curve to an area of consumer behaviour which was subject to considerable variation.

The ordering of variables in the TRRL estimation of saturation level (regression of growth rate on ownership) in preference to the method suggested by Adams was supported by Kirby, with confirmation of Tanner's earlier observations on the increased likelihood of statistical error in the alternative method. It was noted that Adams' method (regression of ownership on growth rate) implied that the current growth rate would be dependent upon the next year's car ownership level. However, Kirby expressed concern over the use of cross-sectional (county) data as the basis for the regression technique used by the TRRL. Cross-sectional data, as used by Tanner since 1962 in the regression of growth rate on car ownership rate, had produced considerable variation in intercept (saturation level). Time-series analysis of national data had produced a more consistent intercept value, usually less than 0.40 cars per person, and had reduced the effect of spatial variation noted previously by Adams. Time-series analysis of different regions, it was noted, suggested spatial variation in saturation estimates, as well as a lower overall figure to that obtained by more recent cross-sectional analyses. In addition, it was suggested that the use of exponential regression could reduce the effects of error

THE IMPACT OF GOVERNMENT INTERVENTION 51

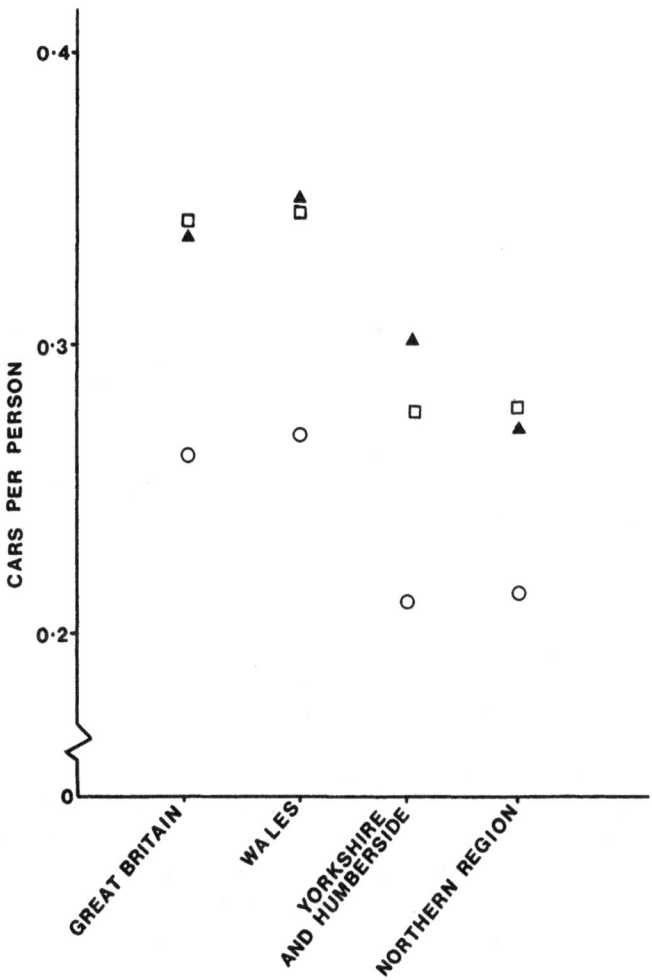

FIGURE 4·3. Estimates of saturation levels for UK regions based on different procedures. (Source: Kirby, 1976.)

structure in variables used to estimate saturation.

Figure 4·3 illustrates the effect of the application of these different techniques to time-series data for selected UK regions (for 1957–1974). Adams' method involved regressing ownership on growth rate; Ioannou's estimates used the standard TRRL procedure, whilst those of Kirby were based on the exponential regression technique.

Kirby's conclusion was that the use of national time-series data, in preference to cross-sectional (county) data, suggested an overall saturation level much lower than that currently in use (0.45). This should be reduced "to a value somewhat less than 0.40". It was noted that the saturation levels for individual regions could still be well in excess of this and that, in general, the higher the current level of ownership, the higher the saturation level was likely to be. In addition, the use of the one simple logistic model, with fixed parameters, was questioned, with the added suggestion that several alternative saturation levels could be used to allow for variations in future levels of income, price of cars and economic activity.

Increased Government Concern

Public concern over national UK transport policy reached a high level in the early 1970's. There was particular emphasis upon the impact of continued development of the primary road system and the means by which central and local government presented the case for future capital spending. Car ownership forecasting techniques, being the major prop upon which traffic growth figures were supported, had by 1976 become an issue of more than academic concern. The popular experience of the effects of the 1973 oil crisis had renewed the debate with a questioning of the assumption of sustained growth and stable fuel supplies. Accordingly, the problem was afforded separate consideration in the parliamentary policy development process.

As part of a consultation document on transport policy (1976), which preceded the publication of a new White Paper, a number of questions of agreed public concern were given an official airing (with particular reference to contemporary Department of the Environment forecasts).

1) Should the forecasts be revised downwards given the current economic position?
2) Is the "saturation level" too high?
3) Do the forecasts take account of possible future changes?
4) Does the Government forecast traffic and build roads to accommodate it?
5) Are the forecasts self-fulfilling (in that roads built to accommodate forecast traffic levels will inevitably generate traffic at that level)?

The last question covered an area of considerable popular and academic concern. Haggett's (1973) statement that "models are to use, not to believe" was developed by Adams (1976) with particular reference to road transport. The same theme was to be expanded by Bennett (1978) in his critique of "black box" planning methods and the employment of "closed loops" (self-fulfilling forecasts) in urban and regional planning.

The initial responses to the five rhetorical questions contained in the consultation document were generally in support of existing policy. Notwithstanding this, the same document gave notice of the intention of the Government "to seek an independent assessment" and to comment on and possibly recommend changes in the existing pattern of trunk road assessment. This was to include an examination of the validity of existing methods of traffic forecasts (including car ownership) and the weights given to environmental and economic factors. An independent body, the Advisory Committee on Trunk Road Assessment (ACTRA) was set up in February 1977 prior to the publication of the June 1977 White Paper on Transport Policy. One of two major terms of reference was "to review the Department's method of traffic forecasting, its application of the forecasts and to comment on the sensitivity of the forecasts to possible policy changes."

The Leitch Report

There is no doubt that the development of mathematical modelling in the field of car ownership has and will be influenced by the findings of the *Report of the Advisory Committee on Trunk Road Assessment* (ACTRA) (the Leitch Report) which was published in January 1978. The fact that a number of recommendations were to form the basis of a Government White Paper, coupled with the lengthy examination of "official" car ownership prediction techniques is, having regard to the magnitude and diversity of previous efforts since 1960, highly debatable. There was evidence that the panel were influenced as much by basic philosophical thinking (about the value of mathematically-based predictions) as they were by a desire to examine the detailed strengths and weaknesses of the methods which have so far been employed. Nevertheless, there was some need for official re-iteration of some of the qualifications which have already and consistently been voiced by those working in the field, but which have been conveniently omitted from the resultant "official" memoranda upon which so much justification for highway planning has been based.

Under the general heading of "uncertainties inherent in the data", the report exposed the many constraints which must be placed upon the use of car ownership prediction figures. Not the least of these is the need to employ caution in the use of official population projections. It goes without saying that, although this may not influence the calculation of saturation levels (a major area of controversy) to a significant extent, it will nevertheless control the actual number of vehicles which are in existence. The Leitch Report noted the wide variations in population predictions which had occurred over the preceding twenty years (see Figure 2·3) and it requires only a simple calculation to show the degree of error in vehicle numbers which would be explained by extremes in forecast. Then there is the question of regional variation in population growth. Similarly, the report noted equally wide uncertainties over energy supplies with the proviso that the immediate future (up to 1990) would most likely represent a period of relative stability in terms of fuel supply and prices, in contrast to the experience of the mid-1970's.

Turning to examination of individual techniques, the report contained extensive criticism of the extrapoaltory "time-trend" model used by the Department of the Environment and which was based on use of time as a proxy for factors determining car ownership which are not included in the model. With justification the report pointed out the necessity for these factors to be stable both historically, and in the future, for the extrapolatory model to work, with the comment that, if these factors are not stable, a causal (explanatory) model is preferable. A new premise here is that the causal model should "adequately represent underlying mechanisms determining behaviour". Despite a great deal of work this, in a nutshell, remains the basic philosophical theme in the car ownership prediction debate.

Because it has such an important bearing upon all predictions beyond the base year, it is understandable that the saturation level concept and its estimation procedure should have come under close scrutiny. A small study of examples of applications of COBA using various saturation levels demonstrated its importance to the fundamental existing method of highway evaluation. Having noted the failure of the TRRL model to take into account changes in income and vehicle costs in the estimation of saturation level, or any allowance for regional variation in saturation level, the report concluded that the method is unsatisfactory. In addition, there is not enough allowance for variations in real incomes, it being noted that the model predicts very little difference in car ownership levels by 2010 for a "no-growth" income situation as opposed to a 2 per cent per annum increase (which would effectively double income by 2010). In addition, the use of the logistic growth curve was criticised both at the level of general criticism of extrapolatory models and with the more central comment that a number of alternative curves were available. As noted earlier, Tanner's choice of the logistic curve was based on initial assumptions of an analogy between motor car ownership and other biological/manufactured phenomena which, with the passage of time, may have been less justified.

It is clear that the Leitch Committee were greatly influenced by the virtues of the econometric, disaggregate modelling which had been the subject of increasing research activity. Not only did this provide a more useful disaggregated form of prediction at household, local and regional level, but it was also seen as providing a causal explanation for changes in ownership. The result was a number of models which, in seeking to offer causal explanation, permitted a deeper understanding of the economic (and social) factors influencing the decision of a household to purchase (and retain) one or more cars. These models require no lesser consideration of the all-important population and energy variables which are central to their final application and validity. Such a model is included in the Regional Highway Traffic Model (RHTM) and it seems likely that this could supersede the TRRL model as the official DTp predictive tool. At the same time, the Leitch Committee noted the need for consideration of family size, structure, and age, which several commentators (e.g. Mogridge) had already cited as critical variables affecting ownership patterns. In conclusion it was suggested that there could be some attempt to merge the predictions based on the RHTM (causal model) with those based on the TRRL method, with the proviso that, since the RHTM represented a more valid procedure, there should be no attempt to modify figures to fit extrapolatory forecasts.

The Regional Highway Traffic Model

Following considerable political and technical discussion, and prior to the establishment of the ACTRA, the Department of Transport had already initiated a major study aimed at improving the use of local data as a basis for projections of traffic growth associated with new highway. There had been a particular problem associated with forecasts of traffic on motorways, partly as a consequence of non-availability of data on origins and destinations of vehicles which could normally be collected by means of roadside interviews. Work on this Regional Highway Traffic Model (RHTM), aimed at replacing an earlier and more aggregated National Traffic Model, commenced in 1976, with a provisional completion date set for early 1978. By that time it was expected that there would be produced a four stage road traffic model. It was planned that four sub-procedures would be developed; car ownership, trip end estimation, trip distribution, and assignment. The study has been beset by problems particularly in the area of calibration of the trip distribution model. In contrast, the development of the car ownership sub-model proved much more successful.

As a result of the interest shown by the ACTRA panel details of the development of the car ownership model were published in advance of the full RHTM results. Its history has been set out by the team responsible for its basic calibration (Bates, Gunn, and Roberts, 1978). Reflecting the philosophy of Bates, outlined earlier, the model assumes that growth in car ownership is a function of two economic variables; real income and the price of cars relative to other variables (in effect, the retail price index). The team are careful to note that, although they have demonstrated a significant relationship between the variables examined (during the period 1961–1975), future levels of car ownership are still dependent upon variations in car prices and incomes, the prediction of which remains surrounded in uncertainty.

Developed as a partially-disaggregated causal alternative to extrapolatory procedures, the final Department of Transport model is based on a level of research input which has been equalled (in the UK) only by the TRRL. Selection and development of the model has been based on objective reasoning in the selection and testing of variables. For a given area the form of the model (1978 version) is shown in the following equations:

$$P_{1+}(I) = \frac{s_1}{1 + \exp[-(a_1 + b_1 \log I)]} \quad (4.1)$$

$$P_{2+/1+}(I) = \frac{s_2}{1 + \exp[-(a_2 + b_2 I)]} \quad (4.2)$$

where P_i = the proportion of households at given income level (I) owning at least i cars
s_1 = saturation level for P_{1+} group
s_2 = saturation level for multi-car owning household group

(Thus $P_{2+/1+}$ = proportion of car owning households owning at least two cars.)

The coefficients a_1, b_1, a_2, b_2, may change according to the location of the relevant zone, thus allowing for variation between separate geographical areas in terms of accessibility. In fact six different "accessibility bands" were identified and employed, each of which produced different values for the coefficients. Accessibility was measured by a simple gravity model which took the form:

$$A_i = \sum_j E_j f(s_{ij}) \quad (4.3)$$

where $f(s_{ij})$ = $0.781 \exp(-0.146 t_{ij})$
E_j = total employment in zone j in 1976

t_{ij} = travel time on 1976 network between zones i and j

j^1 = set of zones lying within 60 minutes travel time from zone i.

Thus the RHTM procedure represents a sophisticated form of model capable of application at a variety of hierarchical levels of spatial zoning. Although consideration was given to the use of an extrapolatory model, the ultimate choice was based on an analysis of households, thus allowing its application to the problem of estimating car ownership for small areas, provided that "the problem of temporal stability could be resolved". This problem, which has proved the most significant since 1978, refers to the likelihood that cross-sectional income/ownership relationships (curves) established at one particular time will remain stable in the future. The category analysis method of trip generation assumes this is so. The concept was tested using Family Expenditure Survey (FES) data for 1969–1975. It was found that, provided income was corrected to allow for changes in the retail price index and car prices, the model remained stable at the national level for a ten year period to an approximation sufficient for medium-term forecasting. There was in fact some evidence of instability over time, particularly during the early 1960's. This was explained by changes in the relative price of second hand cars which, following a sharp reduction in real price during the late 1950's and early 1960's, levelled off towards 1970 and was now relatively stable.

To allow for variations between car prices and the retail price index, income estimates input into the model were "deflated" by use of a car price index. This provided a series of income/ownership probability curves which, assuming stability over time, permitted future estimation of regional levels of household car ownership for a given rate of increase in car purchasing income. (These were similar, in concept, to those shown in Figure 3·3.) In claiming that the model had successfully reproduced the relationship between household car ownership and income observed during the years 1969–1975 it was emphasised that income might not be the sole determinant of variations in ownership and that, within each income band, there might be significant variations from the mean. The latter did represent, after all, a zonal aggregation of households.

THE IMPACT OF GOVERNMENT INTERVENTION 59

The difference between the RHTM model and TRRL procedures was not solely related to concept and form. Several variables and terms were either not used or changed. In addition to the substitution of gross personal income for gross domestic product, the model used car purchase price rather than purchase and use combined (as in the latest TRRL procedures). Of course, the RHTM procedure is a household based model, with no direct reference to personal units. Time, as a variable, is not considered.

Limitations of the RHTM Car Ownership Model

In spite of a successful calibration of the model, using Family Expenditure Survey (FES) data, several trouble areas remained. The model itself, as acknowledged by the research team, was disaggregated only to a partial level. Although it attempts to predict changes in household unit behaviour, it was developed on the basis of national, regional and zonal averages. So the problems of adequate description of within zone variation remain.

Three more problem areas are saturation levels, treatment of multiple car ownership, and the nature of future income distribution.

Saturation levels and treatment of multiple car ownership

In spite of efforts to provide a realistic alternative to earlier national models, the RHTM team were still unable to avoid the issue of saturation levels of car ownership, that is, the old concept that demand for cars should eventually level out (on a cars per head basis). As we have seen, the estimation of saturation values has been a regular source of controversy. The RHTM method involved a subjective estimation of the likely limits to one-car and multi-car ownership in the form of the model parameters s_1 and s_2. Estimation of s_1, the saturation level for single-car ownership, proved the easier task. Initially, it was assumed that almost all households would have access to at least one car. This was later replaced by a 95 per cent saturation level (giving an s_1 value of 0.95) without any significant effect on the validity of the model.

The estimation of s_2, the saturation level for multiple ownership, required more care. An initial value of 0.6 (i.e. 60 per cent of house-

60 CAR OWNERSHIP FORECASTING

holds) was retained with the acknowledgement that a higher value (approaching 0.7) could be justified on the basis of evidence from parts of the United States.

In line with the observations of Ben-Akiva and other workers in the United States the reliability of the model comes under scrutiny in the area of multi-car ownership. Indeed, the team acknowledged this, noting that the effect of changes in household structure could achieve greater significance, but that in the whole area of prediction of multiple-car ownership by high-income households was hindered by lack of existing data.

Estimation of future income distribution

The interim RHTM model contained another limiting factor, namely the assumption on present and future household income distribution. It was accepted that for a single year income approximated to the Gamma distribution with the form

$$p(x) = \frac{a^{n+1}}{\Gamma(n+1)} e^{-ax} x^n \qquad (4.4)$$

where x = household income
  n is a parameter to control shape of the function, and
  a is a parameter controlling scale.

An examination of separate distributions for individual years 1969–1975 showed that the parameter n, which controlled the shape of the function, varied considerably over time and between RHTM zones. Nevertheless, it was considered necessary to apply only one value of n to future years and assume a constant shape to the distribution ($n = 1.23$).

Further criticism of the Department of Transport "causal" model was voiced by Tanner (1978). In addition to the general comments already noted he reiterated the longstanding weakness of household category models related to the assumption that car ownership rates will remain stable over time within a particular income group. It was also noted that there remained a short supply of long range data at the household level and for different areas sufficient to calibrate such models, that estimation of saturation levels and income/price elasticities, noted earlier, was no easier with disaggregated procedures, and that there still remained a need to forecast future income distributions to allow for changes in economic growth, household size and structure. Tanner also questioned the heavy reliance on income growth as the main independent variable and reiterated the view, quoted by many observers, that additional variables were required to describe such factors as changing land use patterns, public transport quality, standard of the highway system, motoring costs, social considerations, and so on.

In spite of problems associated with these factors the interim model was presented with the conclusion that it had been shown to have been relatively stable over a reasonable length of time and that, by using such a model, the growth in car ownership since 1961 could be "entirely attributed" to change in income and relative decline in the real price of vehicles.

Since the publication of the interim model, however, further development has taken place. One major change has been introduced. As a result of testing of more recent family expenditure data, it was found that an expected reduction in car purchasing activity did not occur to the extent consistent with the model. Between 1975 and 1978 car prices in the UK rose in real terms whereas incomes grew slowly. The model prediction for 1978 fell 16 per cent below the actual recorded level. It was considered that this was due to some "time lag" in the impact of real changes in car prices against income and, as a result, the model was adapted to incorporate a lagging effect.

It seemed likely that the RTHM forecasting procedure would replace the TRRL method in official Department of Transport directives to national and local transport planners. Despite the problems related to temporal stability in economic relationships outlined above, the use of a causal model appeared a more appropriate policy than the application of an extrapolatory method based on the theoretical characteristics of a particular form of mathematical curve.

A New TRRL Model

While the RHTM model was under development a new TRRL forecasting procedure was being produced in parallel. This was used as the basis for a new set of forecasts contained in TRRL Report LR 799 (Tanner, 1977). This report proved to be more wide-ranging than any produced previously and demonstrated an attempt to re-examine the criteria for use of particular types of extrapolatory curve with time-series data. Although the procedure for estimation of saturation levels remained largely unaltered, attempts were made to introduce several additional controls. However, possibly in response to a growing awareness of the need for a cross-sectional disaggregated approach, the report also devoted a great deal of space to attempts at improving prediction procedures at the county, and even household level. In addition, there were further attempts to allow for future variations in incomes and fuel costs, even though this could only be done at the expense of jettisoning the concept of a fixed rigid forecast (growth curve), providing instead a series of growth alternatives, and thus placing final interpretation in the hands of others, (others being the officials of the Department of Transport who produce guidance memoranda for national, regional and local authority use).

Because it is claimed that the most recent TRRL forecasting report contains several notable innovations, it is necessary to examine these in reasonable detail to conclude whether these claims are valid or justified. The most notable development is the abandonment of the use of the logistic curve which had formed the basic core of Tanner's work since the early 1960's.

The logistic curve was replaced by a "power growth curve" which, despite variations in algebra, is based on a similar concept and parameters to the one it replaced. Tanner's justification for employing this new curve was that some of the data used were "not consistent with a strictly logistic curve". He appears to be particularly aware of the dangers of extrapolation in use of regression analysis to calculate saturation levels (as had been noted by Adams and others) and the possibility that the relationship used may be influenced by real changes in personal income and motoring costs. An important new consideration appeared to be an acknowledgement that the relationship between percentage growth of car ownership and level of ownership was not linear but curved, and that this curved effect resulted in a slowing down of the rate of growth during the second half of the growth towards saturation. There were "intuitive and theoretical reasons why this should be so".

Although there was no lengthy or substantive explanation as to why the power growth curve was employed in preference to the logistic, much of the previous TRRL analysis was modified or simply re-used in the present analysis, even though a great deal of it was based on the aforementioned assumption of linearity in the relationship used to estimate s (saturation level). The power growth curve took the form

$$y \text{ (cars per head)} = \frac{s}{1 + [a(t-T) + b\log i + c\log p]^{-n}} \quad (4.5)$$

where s, a, b, c, T and n are constants to be determined by calibration
i = Gross Domestic Product (GDP) per person at fixed prices
p = cost of motoring at fixed prices.

In addition to consideration of the two major variables (GDP and motoring cost), Tanner also considered the undoubtedly significant effect of variations in population density in his associated analysis. Although it was acknowledged that a number of studies had shown this variable to be negatively correlated with car ownership, it was noted that there was little understanding of the causative factors involved (e.g. difficulties of parking/garaging, traffic congestion, availability of public transport, length of trips). The basic approach was still to employ an extrapolated curve and to produce a range of forecasts allowing for variation in income and fuel cost at the national level and, in addition, population density at the local level.

The estimation of saturation level (and ultimate control of the growth curve) involved a more complicated procedure than that employed in earlier TRRL work. It was necessary to examine the growth rate/ownership relationship in several ways. To begin with, the conventional long-term national trend gave an expected saturation level of 0.35 but examination of a longer period suggested that the underlying trend was not in fact linear, so that 0.35 provided only a lower bound. Coupled with "experience of other countries" it was suggested that saturation was unlikely to be less than 0.6 cars per person. Further work involved an examination of recent growth rates in different areas and groups to seek to establish levels of ownership at which further growth would have stopped (allowing for the difficulty that no areas

64 CAR OWNERSHIP FORECASTING

or groups have yet reached this stage). Evidence from regression analysis indicated a saturation level of at least 0.6 for Great Britain as a whole. In addition, statistics of the potential driving population were examined and suggested that "except in areas where car owning will be of little benefit" the final level of 0.6 cars per person could be expected.

In conclusion, Tanner was able to sum up that the data as a whole implied a saturation level of about 0.5 cars per person within a range of probability between 0.4 and 0.6. This compared with the 0.45 figure used in previous TRRL studies.

Tanner may have been hinting at the limitations in this forecasting procedure when he reiterated his earlier caveat that levels could only rise beyond 0.5 if land use patterns and transport systems were developed to the extent that they represented significant causal effect and that the levels would not be realised unless economic growth continued and there was some stability in supply and price of fuel.

Allowing for variations in saturation level, income growth, and fuel costs/availability, the final graphical predictions carried forward a stage further the production of alternative growth curves which had begun in the 1974 predictions (see Figure 4·4). However, the margin for variation was much wider, and in common with earlier work, but to an even greater degree, the decision on likely future trends in independent variables (such as income, fuel cost, population density, public transport provision/interaction, government policy) was left to the user. So, too, was the problem of actual population levels (and household structure/driving ability). There is no scientific examination of the likely impact of reduced population levels on the other independent variables (in addition to car ownership rates) and it should be remembered that predictions are in the form of cars per person, not total number of vehicles. Although not directly stated it appears that, despite the provision of high and low alternatives, the philosophy behind this type of extrapolatory prediction begs the acceptance of a "middle" forecast and, in line with earlier policy and procedure, the Department of Transport's *Interim Memorandum* (January 1978) narrowed down the range of forecast in the fashion shown in Table 4·1, while noting that research was still in progress on an official causal model of car ownership (the RHTM).

The problems, noted throughout this book, which have beset the

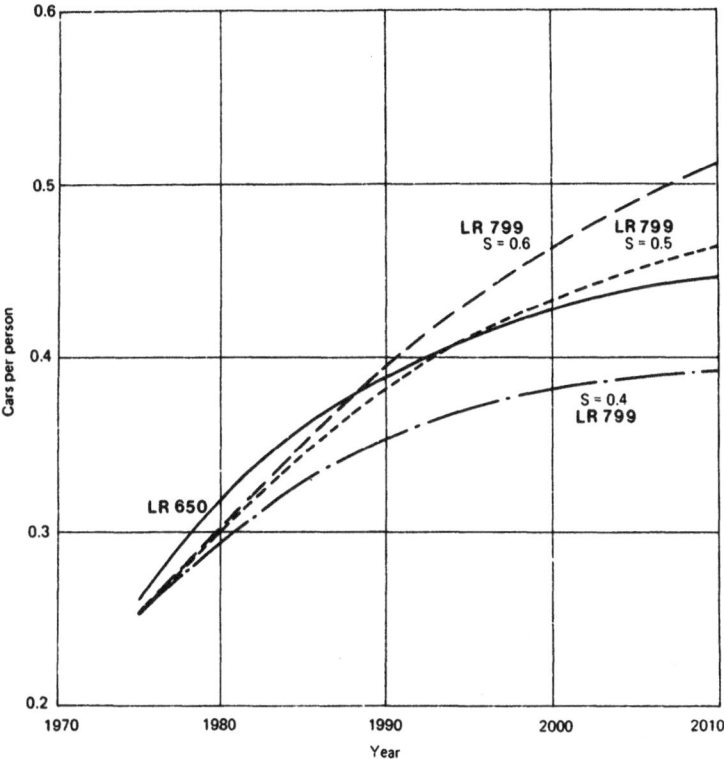

FIGURE 4·4. Comparison between LR650 and LR799 car ownership forecasts. (Source: Tanner, 1977, Figure 1.)

producers of extrapolatory models have been increased by this recent trend of provision of alternative forecasts. The latter appears to provide testimony to the weakness of the use of extrapolatory curves of the logistic and power growth type which assume car ownership growth to be analogous to other consumer goods or biological phenomena. The use of the cars per person/growth rate relationship in the estimation of saturation levels has been consistently questioned and there has been little scope for the use of disaggregated data in the formulation of national models nor the application of nationally based models at local, household or other disaggregated level.

This is not to say that the most recent TRRL procedure did not

TABLE 4.1
D.Tp. INTERIM CAR OWNERSHIP FORECASTS
JANUARY 1978

	Cars per Person		
	Low growth assumptions	High growth assumptions	1975 Forecast (lower)
1976 (Actual)	0.258	0.258	—
1980	0.27	0.30	0.30
1985	0.31	0.35	0.34
1990	0.33	0.39	0.38
1995	0.35	0.42	0.40
2000	0.37	0.44	0.42
2005	0.38	0.46	0.43

make an attempt to introduce a relevant model for use at regional, local (county), and even household level. Regional predictions followed the Kirby method (outlined earlier) with slight variations due to the use of an extended range of time series data. Cross-sectional work involved examination of recent (1971) levels of ownership according to geographical county, household size and socio-economic grouping (of "chief economic supporter" in 1966 or household head in 1971). Analysis of data at personal and household level was merely descriptive and, again, there was no attempt to provide explanation of relationships or the all-necessary future numbers in each group. More detailed analysis was carried out on current ownership patterns at the county level, in an attempt to produce an explanatory model.

This county model took the form of a multiple regression equation to explain cars per person. Using data from the 1971 Census of Population (apart from income) car ownership per person was explained in terms of the following six variables:

1) Income — from an Inland Revenue Survey of Personal Incomes 1970–1972

2) Log of population density — logarithm to base e of persons per hectare in 1971

3) Population growth — per cent increase per year between 1961 and 1971

4) Age 0–14 — the proportion of "under fifteens" in 1971

5) Age 65 + — the proportion of population over 64 in 1971

6) Self employed — the proportion of economically active males in Socio-economic Groups 3 and 12

These six variables were found to explain 89 per cent of the variation in cars per person. Surprisingly, the effect of income was not any more significant than any other variable.

Although this represented an attempt to produce a causal/explanatory model based on the present situation and conditions superseding earlier TRRL local procedures, there was no detailed indication of how it should be used for predictive purposes. As it stands, the use of six independent variables produces a greater scope for error and, in any case, it is clear that some of the variables are themselves correlated. This is the same multi-collinearity experienced by other workers and, despite the inclusion of several different variables, the Tanner model represented no significant advance on the work of Fishwick (see page 41).

The new TRRL national model stimulated renewed debate on philosophy and methodology. Since the substitution of the power growth curve reduced the influence of saturation level on immediate (10–12 years hence) predictions there has been less discussion of the earlier problem of estimation of saturation levels. Instead, several commentators, notably Bates, (letter, January 1978, *Traffic Engineering & Control*) criticised Tanner's assumptions on income elasticity, and economic variables as a whole, and has demonstrated that such is the importance of economic variables that these have a vital effect on saturation levels and, therefore, the entire prediction process.

Tanner (1978) has aired view son the problems associated with the LR 799 model. In reiterating long-stated reservations on the degree of care required in application of extrapolatory forecasts, the following sources of uncertainty have been noted, and qualified, in the following fashion:

Economic growth — Despite short term fluctuations the assumption of moderate growth is justified on the basis of a longer term trend (for the past 50 years in the UK, excluding world wars) of 2.1 per cent per annum.

Fuel prices — Although subject to great uncertainty this was not considered to be of overriding importance. On the basis of UK Department of Energy estimates for oil prices, coupled with refining and taxation considerations, a real increase in retail prices of 25–50 per cent by the end of the century would fall within the range of uncertainty allowed for in the model.

Policy dependence — The TRRL forecasting procedure continues to include the assumption that there will be no "extreme" policies introduced in the field of road traffic.

Self-fulfilment — The claim that new roads induce higher car ownership was noted with the response that any effect of self-fulfilment was likely to be small and was already incorporated (by default) in the forecasting procedure.

Concept of saturation level — It was acknowledged that, since saturation is still a long way off, estimation of the likely level and the nature of car ownership and use at that time is very hypothetical. It was noted that, in the US, there was a clear tendency for vehicle mileages to decrease as car ownership rates increased (owing to lesser use of "second" and "third" household cars). However, in the absence of clear empirical evidence to the contrary, the use of a saturatrion level was justified, if only for convenience as a controlling parameter on the logistic or power growth curve. .

Estimation of intercepts	It was acknowledged that the original method of estimation, using linear extrapolation of the trend of growth rate on ownership level, may not have been completely valid, especially as recent work had shown considerable variation using data on a regional and local basis. The introduction of the power growth curve has compensated for this.

Support for the TRRL method, and criticism of the RHTM procedure has come from several quarters, including Fowkes, Pearman and Button (1979). They reiterated some of the comments of Tanner, adding that explanatory variables may be autocorrelated. This, household income, household size, residential density, and lower public transport accessibility are likely to be interrelated. On the positive side aggregate data for the period 1953–1974 was "reworked" to demonstrate that there was evidence that growth had been logistic. Previous errors in TRRL forecasts were, it was claimed, the result of inaccurate parameter estimations (saturation) rather than use of an inappropriate model. Thus the abandonment of the logistic curve, in favour of power growth, was questioned.

In spite of an earlier (1978) comment that the general development of the RHTM and TRRL methods should continue "until they become identical", in a later (1979) discussion, Tanner added further formal observations on incompatibility between the two. He was sure that there should be more discussion on the role which these models were supposed to perform, that is whether they should be used to give a general, aggregated and "unconditional" projection of future car ownership or whether they should produce more flexible results to allow for different economic and political scenarios as a basis for assisting policy development. It did not always follow that the more complex the procedure the more appropriate was its general application. In such areas as the assessment of individual highway schemes, a simple model, free from the complication of external variables, might be more effective.

Other Econometric Models 1973–1979

Although there is evidence that the latest TRRL forecasts represent an attempt to broaden the base of the problem, it has been left to other parties to attempt to develop the art of cross-sectional and disaggregate analysis during the period since the oil crises of 1973. Once more, economic theorists have devoted a great deal of attention to the problem. Fowkes and Button (1977) noted the need to base car ownership modelling more firmly on economic theory and with "full cognizance of the spatial and temporal dimensions of differences in ownership levels". In particular, the failure to work at local level was noted with reference to the spatial differences in saturation levels previously produced by Kirby, Herrmann and others, with the acknowledgement that there did not "appear to be any soundly based procedure for working from the national to the local level in such matters". In addition, the tendency to obtain good levels of statistical fit using aggregated data, where there is often as much variation within groups as between groups, was noted with particular reference to the study, noted earlier, in the West Riding (Button, 1973). Despite attempts by Herrmann and Kirby (and Ioannou) to produce applicable forecasts at county and regional level, it was noted that . . . "the regions themselves are politically defined and are very nearly as heterogenous in their car ownership characteristics as the country as a whole". Fowkes and Button went on to provide an example of the way in which there can be wide variation between regions which are equal in size (population) but are allocated different saturation levels.

The discussion continued with reflections on ways open to avoid the difficulties of disaggregate modelling. One method suggested is to produce pure time-series projections at the lowest level of aggregation permissible from the data, but it is accepted that it would make individual forecasts unnecessarily susceptible to peculiarities in data for any one area as well as entailing a potentially large loss of information. One important point noted was that there was uncertainty as to whether different regions were at different points in "car ownership time" on a common trend in ownership or whether they were at the same point in time on different but related curves. (This should entail discussion of the real degree of variation between regions and the need, given adequate data, to merely consider application of similar methodology to individual regions in turn without attempts to link them or seek any common ground.)

The proposed answer was a pooled model in which population density is the independent variable used to distinguish regional variations. Population density was used because it was a frequently cited "explanation" of spatial variation in car ownership, being a proxy for differences in the quality of public transport, scale of traffic problem, and in the accessibility of services and employment. (More work is required to establish the degree to which density does actually represent a proxy for these variables, especially in smaller cities and rural/semi-rural districts.) Using regional saturation levels already suggested by Kirby a regression model was produced. Saturation levels (the dependent variable) was significantly correlated (negatively) with two variables. These were simple density (persons per square mile) and degree of metropolitan development (percentage of population living in conurbations greater than 250,000 persons).

Attempts to go on to produce a model based on less aggregated (county) data ended in relative failure (possibly because too few variables were being considered: population density, as shown in earlier work, cannot be used as the sole significant influence on ownership or saturation levels).

Some of the problems associated with disaggregate modelling are summarised by Fowkes and Button (1977) and have already been discussed. Fowkes, Button and Pearman (1979) entered into new discussion of the application of national models at the regional level, with particular reference to inferences in the most recent TRRL Report (LR 799) on the matter. As already noted, they tested various alternative forms of curve to those used by Tanner and concluded that the assumption of a linear trend towards saturation level (in the growth/ownership relationship) was not valid.

This conclusion was shared by Mullen and White (1977) in employing their "new approach" to ownership forecasting. Faced with the need to produce disaggregate forecasts (for Telford New Town) Mullen and White rejected Tanner's use of cross-sectional data (because "it assumes implicitly that all areas are at different points on the same logistic growth curve") and time-series projections (which ignore variations in growth rate caused by economic fluctuations) based on the simplified logistic assumption that overall percentage growth in cars per person (with respect to time) has a fixed linear relationship with car ownership level. It was claimed that this was only true if incomes, motoring costs and other significant variables were each changing at a constant rate of time. As Bates had shown several years earlier, this cannot be assumed. Noting the work of Mogridge in the field of cross-sectional modelling Mullen and White concluded that his model took little account of the effect of household size and structure on car ownership, and also ignored the correlation between income and variables such as levels of employment, and the age and sex of household members, which were conveniently excluded from many models but which are important factors.

In an attempt to avoid the problems resulting from use of logistic time-series curve and disaggregation, Mullen and White introduced a person-based model which included influence of variables such as income and public transport availability. (These, by now, with their proxy, population density, are generally accepted as the major factors explaining spatial variation in car ownership.) Although detailed predictions were produced for a number of "person types" with alternatives allowing for variations in the factors noted above, it is, once more, left to others to predict the actual numbers of persons in each category. The most notable development was the assumption that each group of person types represented a car-purchasing market in its own right and that greater increase in ownership could be expected in, say, female workers, than in male workers. It was noted that in COBA, for instance, it was assumed that business traffic would grow at the same rate as other traffic, yet most potential business car users were already car owners. Mullen and White concluded that second-car ownership was likely to be a crucial factor in ultimate ownership levels and that transport policy (in the form of provision of adequate public transport for housewives) would play a more important role than has been expected by most commentators. In general, their national forecasts tended to confirm the 1974 (TRRL) official figures but they were quick to point out that their assumptions and methods differed.

The problem of disaggregated modelling has also recently been the concern of Downes (1977), in this instance in an attempt to produce a workable model at household level. The result has been the production of an empirically derived model which enables a conversion, either way, between the levels of car ownership for an area and the proportions of households within the area having different levels of car ownership. Although the work has not been concerned with actual prediction of future levels, Downes' claim that the relationship appeared to be stable with "time, geographical area and changes of scale" would suggest that it could form one approach to the problem of prediction at a disaggregated level. A similar household-based model was produced by Saunders and Smith (for the Greater London area) (1977) which also took into account the economic effects of changes in "car purchasing income" amongst households. Thus, this represented another econometric model which related ownership levels to household income and the price structure of the new and used car markets.

While most of the developments in prediction techniques have been going on, Mogridge has been concerned with refinement of some of the econometric assumptions and models noted in earlier sections, and has recorded his views and observations in the form of a number of publications. Following his (1975) rhetorical examination of the question "will car ownership saturate?" many of his recent conclusions and implied philosophy were contained in a "personal view on the future of the car" (1976). He concluded that, even on the evidence of post-1973 developments, there was no sign of a reduction in overall car ownership although there was a distinct trend towards smaller engined cars which consume less petrol per mile (as evident in Figure 4·5). Although new car sales declined as a result of the energy crisis, the total stock of cars in use (registered) continued on a rising trend throughout the period. Mogridge noted that there were likely to be more cars in use than official returns suggested because the non-registration of cars, illegally, was much more likely for the oldest cars in times of economic depression, as it was this section of the total stock for which the licence fee is a large part of total costs. (Recent police press releases suggest that licence evasion remains a serious problem. As many as 12 per cent of all vehicles in regular use are not registered, and therefore not included in official figures.)

At a time when increasing political pressure was being brought to bear in the shape of a pro-public transport lobby, Mogridge continued to predict an increased role for the car in cities and elsewhere. Despite his apparent concern for those individuals who, even at saturation level, will be without regular access to private vehicles, he did not suggest how to provide for them.

What then of the future? We have seen how resilient the car market is to the change in the price of petrol, rapidly switching to smaller cars, just as over the last decade with petrol getting relatively cheaper, cars were getting larger. It seems there is no real reason why the average car should not be forced down in size to at least 1000 cc if not 800 cc. Indeed, motorised vehicles will work well below those sizes too, and petrol consumption will fall correspondingly, if not directly.

The 1980 National Traffic Forecasts

The outturn of this activity on the part of government based and independent workers was the production of a new set of national traffic

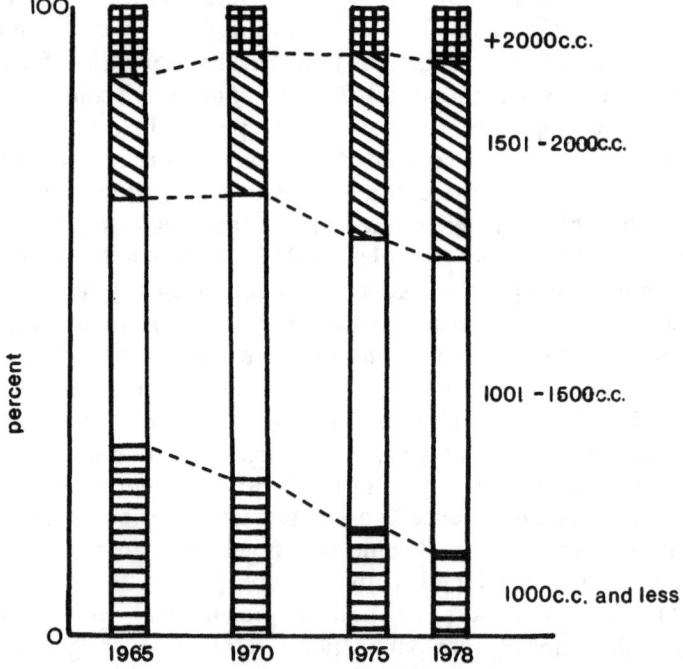

FIGURE 4·5. Changes in the distribution of car engine capacity in the UK 1965–1978. (Source: *Basic Road Statistics*, The British Road Federation, 1979.)

forecasts, published in July 1980, and replacing those contained in the 1978 *Interim Memorandum*. These forecasts were justified in terms of the need to allow for new assumptions on economic growth and further improvement of the causal (household) model to counteract its earlier tendency to underpredict ownership in relation to income growth.

Although primarily intended to provide overall forecasts of vehicle use (road traffic), considerable attention was placed on car ownership prediction. The latter, in anticipation of minor changes in individual vehicle mileages, retains its position as the major perceived determinant of future traffic levels.

Prior to the production of new forecasts, the Department of Transport disaggregate model was subjected to further modification with the inclusion of a "driving licence term" as a proxy variable explaining social desire for improved mobility. It was thought that acquisition of a licence reflected the expectation of future car ownership on the part of the community. Although it was accepted that possession of a licence did not necessarily cause people to buy cars, it was a seen as a proxy expressing changing social attitudes to car ownership which appeared to be an important factor at work but which had "defied precise description and therefore their expression in a mathematical model".

The Department of Transport causal model, as used in the 1980 forecast, now took the form

$$P_{1+}(I) = \frac{s_1}{(1 + \exp[a_1](I/\text{RPI})^{b_1} \text{LPA}^{c_1})} \tag{4.6}$$

$$P_{2+/1+}(I) = \frac{s_2}{(1 + \exp[a_2 + b_2(I/\text{RPI}) + c_2\text{LPA}])} \tag{4.7}$$

where $P_{1+}(I)$ is the probability that a household with gross household income I owns a car or cars

$P_{2+/1+}(I)$ is the probability that a car owning household with gross income I owns two or more cars

s_1 and s_2 are saturation levels (unchanged at 0.95 and 0.6 respectively)

$a_1, b_1, c_1, a_2, b_2, c_2$ are calibration constants

RPI is the Retail Price Index (base 1970 = 100)

LPA is the number of driving licences per adult

Several additional "constants" have been assumed in the operation of the model, namely that the gamma function describing the distribution of gross household income retains its initial shape ($n = 1.6$) and that there is a control on the average number of cars owned by multi-car owning households (2.17 cars per household).

In forecasting driving licences per adult for future years a logistic growth curve is employed using the formula

$$\text{LPA} = s_{\text{LPA}}/(1 + Ae^{Bt}) \tag{4.8}$$

where t is time (0 in 1952)
s_{LPA} is a saturation level (assumed to be 0.8), and
A and B are calibration constants.

The results of runs of the Department of Transport and TRRL (LR 799) models were combined to produce new projections. On a graphical basis, two lines were produced, the TRRL model results being based on low growth high fuel cost assumptions. A line drawn between the two was taken as a lower bound to final forecasts of overall traffic flow. The upper bound is based on a faster rate of income growth, a higher saturation level, and less pessimistic assumptions on real cost of fuel. Official forecasts of car ownership growth are set out in Figure 4·6.

The methodology behind the 1980 forecast contains an interesting paradox in terms of an attempt to achieve some compatibility between two conceptually different models. On the one hand the TRRL extrapolatory model has been considerably modified over the years, in an attempt to make it more accountable to policy or economic variables which have been shown to be significant in cross-section and over time. Nevertheless, the model still involves the acceptance and estimation of a realistic saturation level to be achieved at some specified future data.

Although termed a household model, the complementary Department of Transport procedure is aggregated to the extent that it uses zonal and regional average values for certain parameters. In addition, three saturation terms are required, all of which lie outside the range of present UK general experience. Whilst the assumed value of s_1 is unlikely to be questioned, the corresponding estimates of s_2 and s_{LPA}

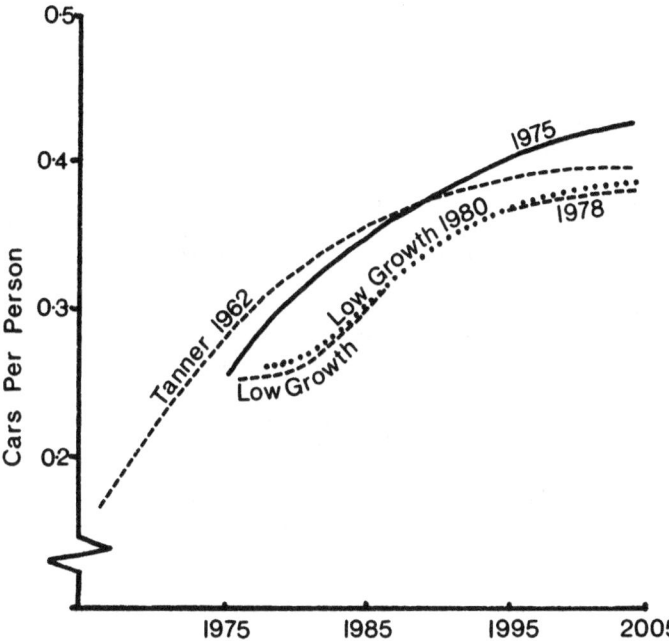

FIGURE 4·6. A comparison between Department of Transport 1980 car ownership forecasts and earlier projections.

could trigger the type of discussion which led to review of the earlier TRRL methods.

The problems surrounding further development of disaggregate modelling, particularly in smaller areas (regions, counties, individual towns) could be exacerbated by the official status of the 1980 forecasts. It is Department of Transport policy that local forecasting of car ownership or road traffic should conform to the national forecasts, even though they may be based on survey work at a fine level of disaggregation. In the meantime it is acknowledged that work will continue at national level to further refine causal models.

REFERENCES

Adams, J., Saturation planning, *Town and Country Planning*, December 1974, pp. 550–554.

Adams, J., Saturation planning — a letter, *Town and Country Planning*, April 1975, pp. 214–215.

Bates, J.J., Gunn, H., and Roberts, M., *A Disaggregate Model of Household Car Ownership*, DOE/DTp Research Report No. 20, London, HMSO, 1978.

Bennett, R.J., Forecasting in urban and regional planning closed loops: the examples of road and air traffic forecasts, *Environment and Planning*, X, No. 2, 1978, pp. 137–144.

Bonus, H., Quasi-Engel curves, diffusion, and the ownership of major consumer durables, *Journal of Political Economy*, 1973, pp. 655–677.

Button, K.J., Motor car ownership in the West Riding of Yorkshire: some findings, *Traffic Engineering & Control*, 14, 1973, pp. 76–78.

Button, K.J., Fowkes, A.S., and Pearman, A.D., Modelling regional and national levels of car ownership, *PTRC Seminar Proceedings, Transportation Models*, PTRC, London, 1977, pp. 33–49.

Department of the Environment, Standard forecasts of vehicles and traffic, *DOE Technical Memorandum H3/75*, HMSO, London, 1975.

Department of Environment, *Transport Policy: A Consultation Document*, 2 vols., HMSO, London, 1976.

Department of Transport, National traffic forecasts, *DTp Interim Memorandum*, HMSO, London, 1978.

Department of Transport, *Transport Policy*, HMSO, London, 1977.

Department of Transport, *Report of the Advisory Committee on Trunk Road Assessment*, HMSO, London, 1978.

Department of Transport, *Forecasting Traffic on trunk Roads: A Report on the Regional Highway Traffic Model Project*, HMSO, London, 1980.

Department of Transport, *National Road Traffic Forecasts*, HMSO, London, July 1980.

Dowhes, J.D., The distribution of household car ownership, *PTRC Seminar Proceedings, Transportation Models*, PTRC, London, 1977, pp. 24–32.

Fowkes, A.S., Button, K.J., and Pearman, A.D., An appraisal of aggregate and disaggregate car ownership forecasting methods, University of Leeds Institute for Transport Studies, Working Paper No. 110, 1979.

Haggett, P., *Geography: A Modern Synthesis*, Prentice-Hall, New Jersey, 1973.

Ioannou, A.C., Regional car ownership forecasting, Unpublished M.Phil. Thesis, University of London, 1975.

Kirby, H.R., The saturation level of car ownership: estimation problems and a regional time-series analysis, *PTRC Seminar Proceedings, Urban Traffic Models*, PTRC, London, 1976.

Kirby, H.R., The logistic model of car ownership: analysis corresponding to its error structure, Unpublished paper presented at the eighth Annual Conference of the Universities Transport Study Group, January 1976.

Mogridge, M.J.H., A personal view on the future of the car, *Traffic Engineering & Control*, 17, 1976, pp. 101–115.

Mullen, P., and White, M., Forecasting car ownership: a new approach, published in two parts in *Traffic Engineering & Control*, 18, 1977, pp. 354–361, 422–426.

Saunders, M.B., and Smith, J.E.R., The GLTS car ownership forecasting model, *PTRC Seminar Proceedings, Transport Models*, PTRC, London, 1977, pp. 12–33.

Tanner, J.C., Forecasts of vehicles and traffic in Great Britain, *Transport and Road Research Laboratory Report LR 650*, Transport and Road Research Laboratory, Crowthorne, 1974.

Tanner, J.C., Car ownership trends and forecasts, *Transport and Road Research Laboratory Report LR 799*, Transport and Road Research Laboratory, Crowthorne, 1977.

Tanner, J.C., Long term forecasting of vehicle ownership and road traffic, *Journal of the Royal Statistical Society*, (A), 1978, pp. 14–63.

Tanner, J.C., Choice of model structure for car ownership forecasting, *Transport and Road Research Laboratory Supplementary Report 523*, Transport and Road Research Laboratory, Crowthorne, 1979.

5 A Review of US Modelling Prior to 1962

CAR OWNERSHIP forecasters working in the United States have followed a different course from their British counterparts. Not only has the country experienced a different pattern of growth in ownership, but there has been a greater diversity in type of authority, advisory body and academic institution possessing an interest in the field. This has led to a great number of modelling applications over the years without the strong constraint of a central controlling authority such as the United Kingdom Department of Transport.

Simple National Procedures

Early attempts at car ownership prediction at national, regional and local level tended to exhibit the same crude characteristics as their European counterparts, with a similar heavy emphasis upon national extrapolatory projections. These were usually produced in the context of the provision of base data for input into comparatively sophisticated traffic engineering models. So we have the US *Traffic Engineering Handbook* (1950), a heavily-circulated compendium of traffic *engineering* procedures, containing virtually no guidance or discussion on car ownership prediction, other than an assumption that past trends will continue. This is in spite of a lengthy discussion on trip generation, and modal split, related to the size of urban areas.

By the time the third edition of the handbook had appeared in 1965 there had been a great deal of external discussion on car ownership growth. However, the manual still contained few details on prediction measures, save for the inclusion of several graphs illustrating past trends in various parts of the country. This was coupled with an implicit understanding that extrapolation of these trends was to be used as the basis for estimation of factors for zonal traffic increase. These factors were to be estimated by multiplying the average annual increase in population, vehicle registrations and use (mileage). As in the case of most simple traffic growth factor models, there was no attempt to allow for a possible saturation point in car ownership or to establish a more flexible causal relationship between car ownership and some other socio-economic variable such as household, or national, income.

At the time of the publication of the first handbook, the early 1950's, some attention was also being paid to car ownership prediction at the macro (national) level. In 1952 the *President's Materials Commission* produced some projections on the basis of an extrapolation of past trends which contained a national prediction for 1975 of 85 million vehicles in use (in reality, the 1975 figure for total vehicles was 120.497 million vehicles, of which 95.241 million were automobiles).

It has already been noted that, during the period preceding the commencement of extensive regional transportation studies, the national emphasis was almost entirely concerned with car ownership rates as an input into zonal traffic growth models for use in traffic engineering. However, the Eno Foundation's *Highway Traffic Estimation* (Schmidt and Campbell, 1956) did contain some discussion of two themes which were to figure prominently in later British work, namely the concept of a "saturation level" for national vehicle ownership and, secondly, an appreciation of the apparent relationship between vehicle use (and ownership) and national income growth.

In spite of a heavy emphasis on "traffic growth" rates rather than car onwership, this manual does provide a convenient reference point from which to measure the degree of progress in predictive modelling which had been made up to this point. Several avenues were examined. On a very superficial basis, there was some discussion of the degree of correlation between ownership, traffic growth, and national income. In addition, the shape of alternative curves of traffic growth was analysed, with particular reference to the concept of a non-linear trend (see Figure 5·1). This provides a contrast with earlier projections and advice which had been based on a simple linear extrapolation of post-war trends. It is interesting to note that at this stage the preference for a non-linear projection procedure took the form of a compound growth curve based on the rate of increase which the

82 CAR OWNERSHIP FORECASTING

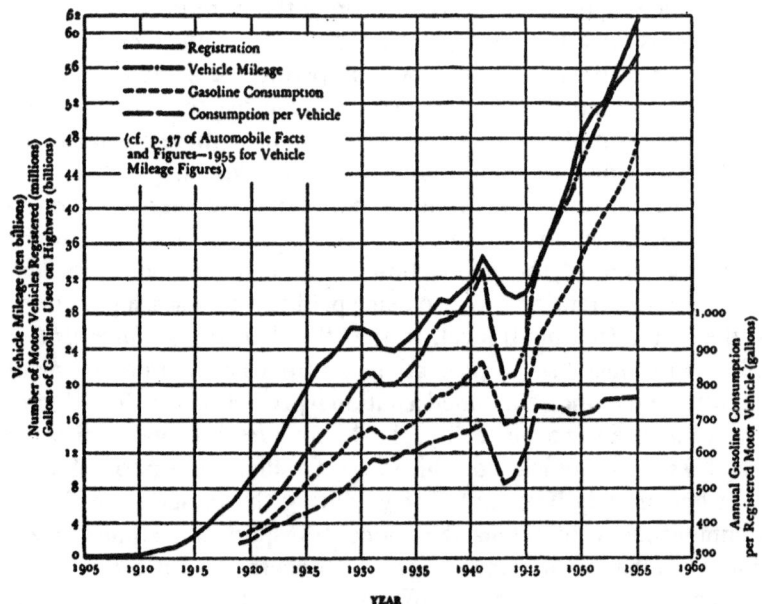

FIGURE 5·1. United States trend in motor-vehicle registration, gasoline consumption, and travel, 1905–1955. (Source: Schmidt and Campbell, 1956, Figure V-3.).

United States had experienced in the early 1950's. This involved the acceptance of an increase in the rate of growth entailed in a compound curve. A consequence of this was the need to consider whether or not the rate of growth would eventually decline as the move towards a "saturation point" began to take effect. It appears that this concept was recognised in the longer term but was considered less relevant to the task of producing interim growth factors for inclusion in local expeditious traffic engineering models.

Nevertheless, the Eno manual did introduce a longer term view of car ownership and the effect of saturation (at a national level) by examining the nature of the *ratio* between persons and vehicles. When this was examined for the period 1940–54 it was found that the ratio of vehicles to persons had displayed a tendency to reduce at an increasingly smaller rate. The average annual rate of increase between 1940 and

1950 had been 2.53 per cent — between 1950 and 1954 it had been 2.12 per cent.

In some ways the complexity and inconsistency inherent in the car ownership modelling problem are evident here. A recommendation that overall traffic growth projections be produced here by the application of compound growth curves is countered by an acceptance that the growth rate of personal and household vehicle ownership may already be beginning to show signs of reduction. Here, however, part of the explanation of the reduction in person/vehicle ratio is seen in a corresponding fall in average size of household, at a time when the overall national population was still increasing rapidly.

The problem of estimation of the year and level of vehicle ownership saturation, to be vigorously discussed in Britain after 1962, was already causing some concern in the United States during the 1950's. Schmidt and Campbell considered a long range national level of 0.5 cars per person to be realistic, noting that "as to the saturation point and its date, one can only surmise". It was recorded that California presently registered a passenger car per 2.4 residents.

On the basis of a combination of these assumptions and procedures a projected national ownership level for 1975 of 2.0–2.2 persons per vehicle was produced, which, when combined with US Census Bureau forecasts of population (medium growth values) gave a vehicular population of 100 million. Comparison with actual 1975 values show that even this figure proved to be a considerable underestimate.

The use of national extrapolatory curves continued as the main method of prediction of vehicle numbers throughout the 1950's. Todd (1960) undertook an extensive analysis of graphical trends at national, regional and state level in order to produce estimates of vehicle ownership, and fuel consumption up to 1976, together with travel mileage up to 1991. No complex mathematical procedures were used, merely a simple linear extrapolation of 1946–1956 trend lines. Final national projections of total vehicles in use were closely linked to population projections produced from other sources (see Figure 5·2).

Even so, this simple method produced an acceptance of the concept of a saturation level almost by default. The 1976 estimate of the person/vehicle ratio was two people to one vehicle (0.495 vehicles per person). Although there was no comment on saturation levels, an examination of Figure 5·2 indicates a reduction in the *rate* of growth of

84 CAR OWNERSHIP FORECASTING

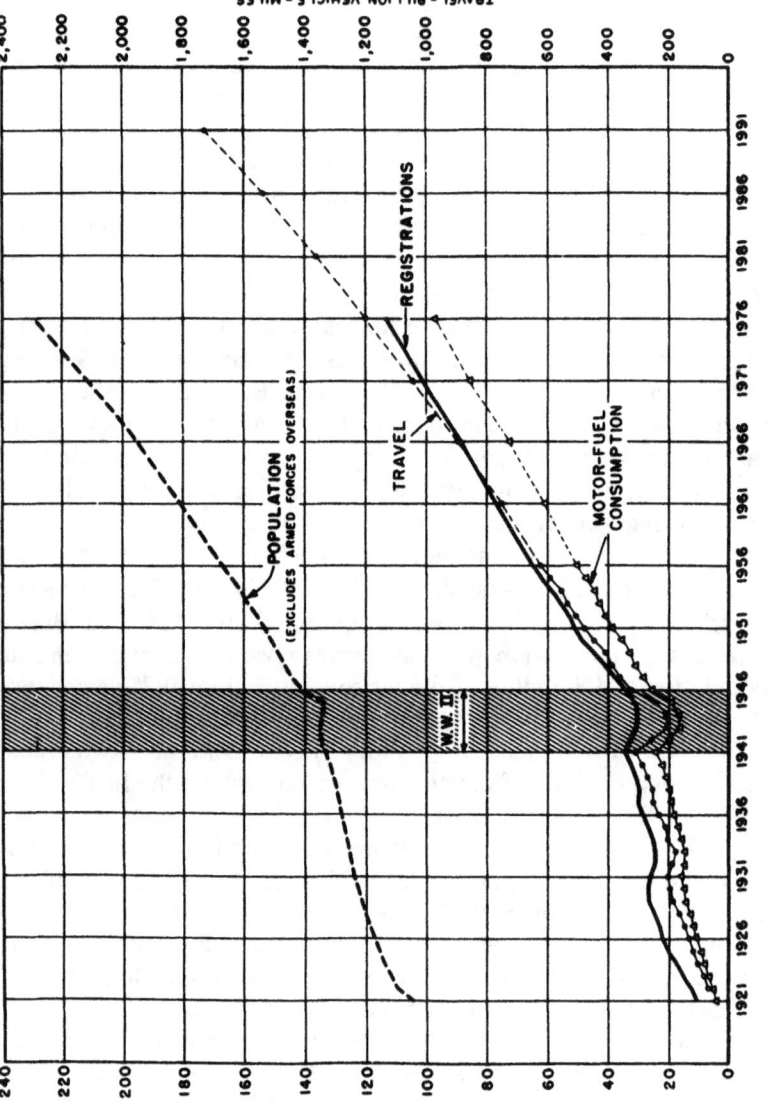

FIGURE 5.2. US automobile ownership trends and projections. (Source: Todd, 1960, p. 262.)

vehicles compared to population in line with the earlier work of Schmidt and Campbell.

Car Ownership and Land Use/Transportation Studies

Following conception in the early 1950's, the application of land use/transportation modelling procedures began in earnest in the later years of the decade. Vehicle ownership rates were accepted as being a major variable influencing the nature and scale of traffic generation. At an early stage of development of land use/transportation planning it became clear that some method of car ownership prediction, other than simple linear extrapolation, was required if this critical element was to be effectively modelled at regional and area level. The *Chicago Area Transportation Study* (CATS) (1960) was typical of the earlier generation of land use/transport plans and is therefore worthy of closer examination in the context of vehicle ownership forecasting. The now-familiar transport planning methodology was employed; base year transport and socio-economic surveys were conducted and the results analysed to produce predictive models of future land use and transportation activity. In this case, a 1956 base provided data for prediction to a 1980 final design year.

As an early input into the entire transport modelling procedure, car ownership data was subjected to an analysis which was much more rigorous than the national work noted earlier. Particular emphasis was placed on an observed relationship between household income and ownership. In addition, there was some allowance given to variation in the spatial distribution of automobiles, using residential density values as a proxy for factors which were difficult to measure.

When compared to more recent applications of modelling, the CATS procedures appear rather crude. Despite a need to predict zonal values for income and vehicles, the correlation between the two variables and the estimates produced were in fact based on, and constrained by, area-wide and national data based on the work of Todd (see Figure 5.3). Thus, the area-wide forecasts for the design year were not based primarily upon an aggregation of zonal estimates, but on a scaling down of these national forecasts to allow for certain land use and socio-economic characteristics which were observed to have a

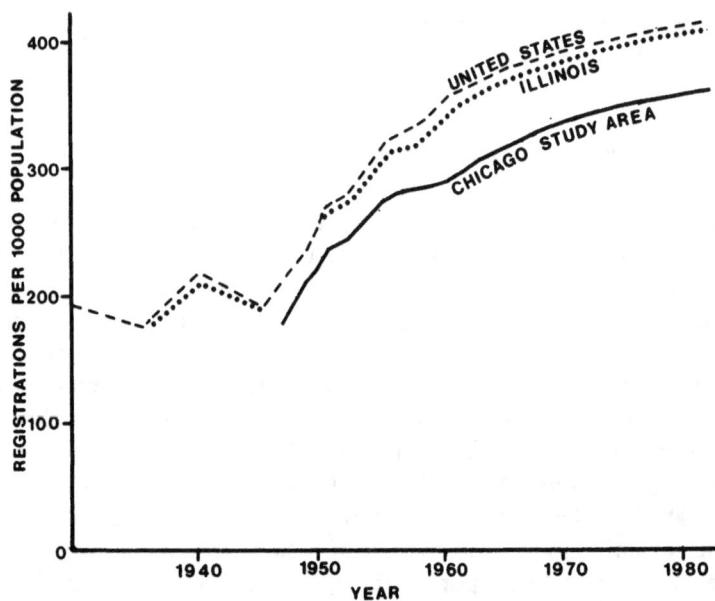

FIGURE 5·3. Automobile registrations per 1,000 population in the Chicago study area, 1930–1959 and estimated to 1980. (Source: *Chicago Area Transportation Study*, Volume II, Figure 5.)

negative effect on ownership. It is of interest to note that, when the national projections were applied in this area study, the concept of a saturation level was introduced (at 0.5 cars per person).

Allowing for these national and area constraints, projected automobile ownership was established by relating ownership to household income. Part of the separate economic forecast had projected the number of households in each income group and, following this, the percentage of each income group owning one and two or more cars (in 1956) was established on an area basis. Inherent in this procedure was, of course, the assumption that as households moved into successively higher income groups their pattern of car ownership would reflect this. In addition, a factor was included to allow for an historical observation that the overall consumer expenditure on transportation had increased, but that the relationship between car ownership and income (at one point in time) was curvilinear. It was observed that, in

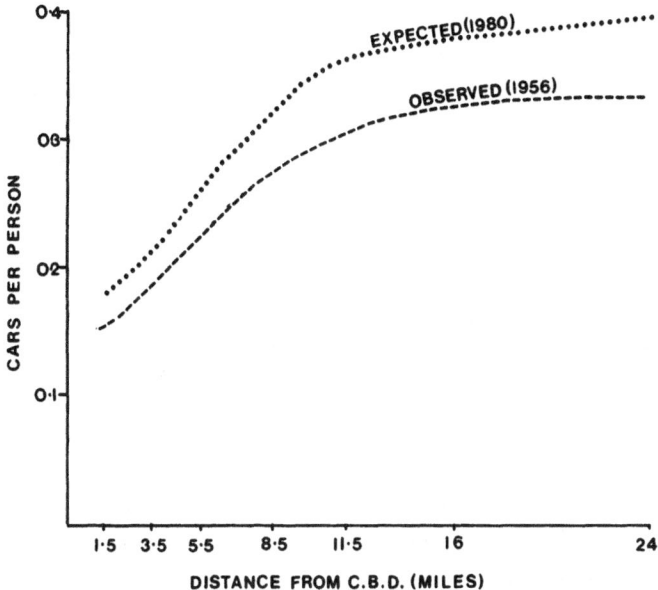

FIGURE 5·4. The relationship between car ownership and distance from the central city area in Chicago. (Source: *Chicago Area Transportation Study*, Volume II, p. 38.)

historical terms, continually rising incomes in the more affluent groups had a progressively smaller effect on car purchase. Thus, the rate of car ownership growth was seen to be highest in only moderately affluent areas of the city. The overall effect was similar to the pattern shortly to be propounded by Tanner in the UK. Its spatial expression is shown in Figure 5·4 which compares relative expected rates of increase on the basis of distance of residence from the CBD.

Following on from this, the study team produced zonal and areawide estimates of car ownership for 1980 together with associated projections of traffic generation, distribution and modal split. The number of autos was expected to double by 1980. The recommended network of primary roads would, if constructed, go towards ensuring that the car ownership forecasts were realised in self-fulfilling manner by "doubling of the demand for highway travel".

Some later work has exposed the weaknesses associated with the use of cross-sectional income/ownership relationships and the general reliance of zonal averages with their tendency to conceal intra-zonal variation. However, the CATS study together with others dating from the same period, set up a basic methodology which is still accepted as the main form of approach to area-wide transport planning in many developed countries. A major innovatory feature appeared with the Puget Sound study (1964) in the form of the application of the zonal cross-categorization trip generation method. In an attempt to achieve more reliable base year data and future year projections at the zonal level the study team introduced the now familiar form of category analysis. Households in each zone were allocated to the economic/demographic categories on the basis of car ownership, family structure and median income of the head of the household. In spite of a wide-ranging academic debate on the behavioural weaknesses of the method, the category analysis procedure, with its use of car ownership/income cross-sectional probability curves, became a popular method of trip generation prediction in the USA, and in many other developed countries where standard regional transport/land use studies were conducted.

Away from the practical necessity of developing car ownership models for use in statutory regional transportation studies, an academic debate on the issue had already been developed. This was primarily centred on the USA, although comparable work in other developed countries is noted in the literature where relevant. At the basic level, the most popular definitive text of the period (Oi and Shuldiner, 1962) affords great attention to the importance of car ownership as a trip generation and modal split primary variable. Oddly, this work contains no discussion of prediction procedures and, in spite of a heavy emphasis on the development of multiple regression equations for trip generation, the problem of stability of car ownership coefficients over time (into the future) is not highlighted.

At a more academic level, a major reference point was the general summary of prediction techniques produced by Martin, Memmott and Bone (1961). Three possible methods for estimating future vehicle ownership was presented: budget study, time series, and a combination of the two. Budget study was summarised in terms of the production of a graphical representation of the household income/car ownership relationship for the region under consideration (see Figure 5·5). This contained a reservation that the assumption of stability in the re-

A REVIEW OF US MODELLING PRIOR TO 1962

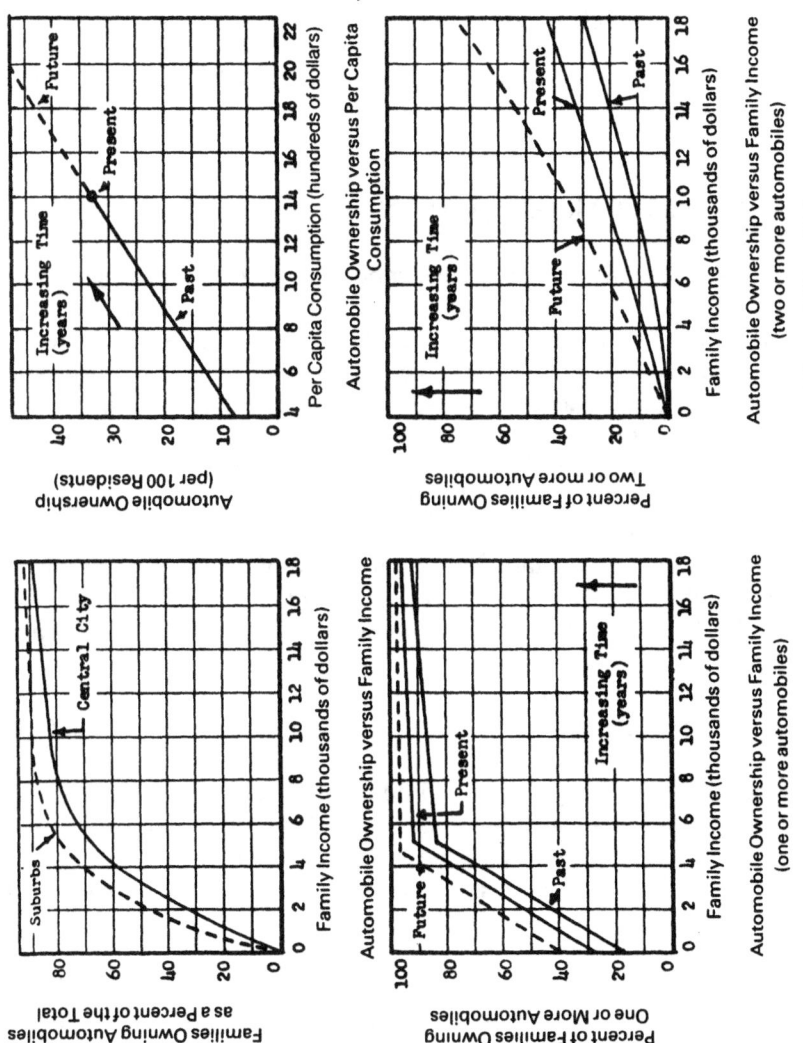

FIGURE 5-5. (Source: Martin, Memmott, and Bone, 1961, p. 73.)

lationship over time may not be valid. The second method discussed, time series, involved a straightforward extrapolation of present trends using a linear regression equation relating car ownership to per capita income. Although the concept of saturation was noted (at a possible rate of 0.5 cars per person), there was no discussion on how this could be built into the equation to produce an eventual levelling-off of the demand curve.

US Econometric Modelling 1938–1962

Partly in response to the relative importance of the automobile industry in the United States there has been a long history of the development of econometric models. Since the 1930's, in academic, commercial and administrative circles, the story has been one of development and testing of economic models, usually employing either time-series or cross-sectional aggregate (national) data.

Prior to the Second World War the academic debate had developed as a result of work by de Wolff (1938). Using time-series national data for the period 1921–34 he examined the relationship between new car production and demand. This entailed use of regression analysis as well as the logistic growth (to be used later by the TRRL in the UK). De Wolff broke down the car market annual change in the following terms:

$$P = S + \Delta R. \tag{5.1}$$

where P = yearly auto production
S = number of cars needed to replace scrapped cars — the "demand for replacement"
ΔR = number of cars increasing the present number of cars — the "demand for first purchase"

The curve of demand for first purchase was assumed to exhibit a logistic shape controlled by three parameters — a saturation value, a "maximum buying velocity", and a moment at which the maximum is reached. Relating the pattern of scrappings/productions between 1921 and 1934 to two variables (car prices, corporation profits), de Wolff developed a regression equation to obtain elasticities of automobile demand. Thus, provided assumptions were made regarding the future pattern of prices and profits, it was claimed that future estimates of demand could be produced, although de Wolff did not attempt any projection.

In a criticism of this analysis, Solo (1939) introduced a number of additional considerations, particularly the need to look beyond absolute numbers of vehicles and examine the size characteristics of the national stock of cars. A re-examination of the data for 1921–34 revealed the elasticity of demand to decrease in times of prosperity and increase in times of depression, this being explained by the price structure of the automobile market. With the depression there was an increase in the proportion of smaller cars sold; the economic recovery saw an increase in average engine size.

In the same year, Roos and von Szeliski (1939) took the analysis further by dividing national demand into two parts, new owner sales and replacement sales, with the two markets producing different coefficients for independent variables.

Academic work continued in similar style after the Second World War, with the emphasis on use of national economic data to produce reliable price elasticities of demand for vehicles. Often, the motor car was perceived as a typical durable good upon which general consumer demand models could be drawn. Atkinson (1952) produced further refinements of earlier elasticity studies, but Farrell (1954) attempted to produce a predictive model capable of direct application. This involved the use of cross-sectional US data (from a 1941 national survey of income and car ownership) to develop an econometric model relating demand to three variables: income, price, and "various taste factors". In addition to establishing a demand function Farrell identified a supply function based on income, price and "other variables related to the supply of cars". The predictive accuracy of the model was tested using 1947–1953 comparative data. Some unexpected variations occurred. In the light of more recent problems encountered in the field of temporal stability of income/ownership relationships developed from cross-sectional data, it is interesting to note the explanation that the success of the model lay in the assumption that the data reflected an unusual "steadily expanding economy" which could be regarded as an "intermediate cause". In a more typical "stationary economic state", it was argued, demand would be likely to depend more on changes in tastes or designs rather than changes in incomes, stocks and prices, and the model would be "relatively useless".

Bandeen (1957) attempted to refine existing methodology and introduced new variables into the discussion. A major development was the examination of spatial variations in automobile "consumption". Using state cross-sectional data and multiple regression analysis (for the period 1940–1950) he found an average income sensitivity of automobile consumption to be 0.9, that is . . .

a 10 per cent change in income was associated with a 9 per cent change in automobile consumption in the same direction. However, an equally significant finding was that a 10 per cent difference in state population density was associated with a 3 per cent difference in consumption in the opposite direction. Consequently, the income/consumption relationship was found to be least significant in states exhibiting high densities of population.

Further explanatory variables were introduced into analyses. Using, once more, national economic data, Suits (1958) demonstrated that over the period 1929–1956 income elasticity had been significantly greater than elasticity of price, with the most rational explanation being the particular history of credit facilities (introducing the concept of "car-purchasing income" used by later UK workers). Following criticism by Nerlove (1960), the Suits model was refined (1961) by introduction of alternative measures of real income. None of this work involved any prediction, and there was heavy emphasis on national aggregate figures. Chow (1957) looked at the value of the national stocks of cars in use and his "stock adjustment model" identified a demand for all automobile services (running costs, purchase costs) as a function of income, relative prices and other variables. This concept was further developed by Frechtling (1963) and Briefs (1963) and has been extensively modified in more recent national models.

REFERENCES

Atkinson, L.J., Consumer markets for durable goods, *Survey of Current Business*, US Department of Commerce, 32 (19), April, 1952, pp. 10–24.
Bandeen, R.A., Automobile consumption, 1940–1950, *Econometrica*, 1957, pp. 239–248.

A REVIEW OF US MODELLING PRIOR TO 1962

Briefs, G.E., Developing auto customer profiles, *Dynamic Aspects of Consumer Behaviour*, Foundation for Research on Human Behavior, Ann Arbor, Michigan, 1963.

Chicago Area Transportation Study (CATS), Volume II, Chicago, Illinois, July 1960.

Chow, G.C., *Demand for Automobiles in the United States*, North Holland, Amsterdam, 1957.

Cope, E.M., and Mundy, A.R., 139 million drivers in 1980, *Public Roads*, XXXIII, October 1964, pp. 68–79.

Farrell, M.J., The demand for motor cars in the United States, *Journal of the Royal Statistical Society* (A), 1954, pp. 171–193.

Frechtling, J.A., A long run look at automobile demand, Proceedings of the University of Michigan Conference on the Economic Outlook, Michigan, October 1963.

Hoch, I., Autoregistration forecasts revisited, *CATS Research News*, III, 1959, No. 2.

Institute of Traffic Engineers, *Traffic Engineering Handbook*, Third Edition, Washington, D.C., 1965.

Martin, B.V., Memmott, F.W., and Bone, A.J., *Principles and Techniques of Predicting Future Demand for Urban Area Transportation*, MIT Research Report No. 38, Cambridge, Massachusetts, 1961.

Nerlove, M., Professor Suits on automobile demand, *Review of Economics and Statistics*, XXII, February, 1960, pp. 102–105.

Oi, W.Y., and Shuldiner, P.W., *An Analysis of Urban Travel Demands*, Northwestern University Press, Evanston, Illinois, 1962.

Resources for Freedom, US President's Materials Policy Commission, The Paley Report, Washington, D.C., 1952.

Roos, C.F., and von Szeliski, V., Factors governing changes in domestic automobile demand, in *The Dynamics of Automobile Demand*, General Motors Corporation, New York, 1939.

Schmidt, R.E., and Campbell, M.E., *Highway Traffic Estimation*, Eno Foundation, Saugatuck, 1956.

Solo, R., The demand for passenger cars in the United States, *Econometrica*, July 1939, pp. 271–276.

Suits, D., The demand for new automobiles in the USA 1929–1956, *Review of Economics and Statistics*, XL, 1958, pp. 273–280.

Suits, D., Exploring alternative formulations of automobile demand, *Review of Economics and Statistics*, XLIII, 1961, pp. 66–69.

Todd, T.R., Forecasts of population, motor-vehicle registrations, travel, and fuel consumption, *Public Roads*, XXX, 1960, pp. 261–274.

de Wolff, P., The demand for passenger cars in the United States, *Econometrica*, April, 1938, pp. 113–129.

6 Recent Developments in the USA and Australia

THE LEVEL of academic and commercial resources applied to demand modelling in the United States far outstrips that experienced in any other country. An extensive programme of technical analysis and model development has occurred over the past twenty years, with a corresponding production of supportive literature. It lies beyond the scope of this book to attempt a minutely detailed commentary on every aspect of this work. Accordingly, this chapter is not exhaustive, but contains details of the more significant contributions in the field.

In comparison to the United Kingdom, work in the prediction field is not only marked by the continued high level of activity but also the significant role played by commercial interests, particularly automobile producers and economic consultants. Here, to a greater degree than elsewhere, the accent is on analysing and estimating the level of consumer demand for cars. The use of prediction procedures for highway planning purposes forms a separate activity.

Deutschman (1967), reviewing the development of national and area modelling, noted the limited scope of much of the work conducted up to the mid 1960's. As was demonstrated in the context of the CATS project, the variables used most commonly were household structure, residential density and, most importantly, mean household income. There was a clear tendency, again evident in the Chicago study, for zonal aggregated projections to be subject to an area-wide control total achieved by simple extrapolation of time series trend data on automobile registrations.

The period since 1962 has seen a number of changes in the pattern of academic research into modelling. Although some extrapolatory models have been developed, the accent has been upon the testing of increasingly sophisticated econometric (causal) procedures. Several observers have commented on this trend (Ben-Akiva 1976, Train 1979) and have developed a form of classification. It is intended to follow standard practice and to examine models in three main groups; these are area non-causal extrapolation, aggregate econometric and disaggregate econometric.

Area Extrapolation Procedures

Although the widespread development of naive extrapolatory models has been overtaken by more sophisticated and accountable procedures, several notable models have been produced in recent years. As part of a general review of car ownership modelling Deutschman (1967) used this technique to produce regional and national forecasts. Trend analysis also formed the basis of more complex procedures produced by Whorf (1973), aimed at forecasting the national pattern of ownership over a lengthy time period and allowing for progress towards national saturation. These were all based on past trends of a number of variables, including household car ownership, house type and house tenure in relation to ownership, car use and scrappage, as well as implied saturation in number of automobiles per licensed driver. Household car ownership was represented by a logistic function of the form

$$F_i = k_i[1 + \exp(A_i - B_i t)] \tag{6.1}$$

where i = ownership class 1, 2 or 3 (cars per household)
k_i = saturation values, $0 < k_i < 1$
A_i and B_i = parameters for ownership class i
t = year

Allowing for additional variations in gasoline prices and car size distribution projections of car sales 1973–1985 were produced.

Although no projections have been produced, the pattern of growth in the USA has been used as a base for the development of models by workers at the UK Transport and Road Research Laboratory. Tanner (1977) reported on the examination of two main aspects of US experience. These were:

1) National trends in car ownership over time with particular reference to income and GDP levels, and evidence of saturation.
2) Empirical studies of growth in areas of high population density (particularly cities). These have been carried out to discover signs of a lower saturation point in urban areas as a prerequisite to the UK model, to allow for high residential densities.

Urban areas exhibiting a lower rate of growth have been subjected to regression analysis and this has exposed a much weaker ownership/income relationship than that evident nationally. In addition, it has been noted that in some cities income growth is lower than the national average, with a corresponding negative effect on car ownership. (Although some cities experienced little or no growth in car ownership between 1960 and 1970 it must be noted that this is partly a result of residential and industrial relocation processes which were not matched by corresponding changes in city boundaries and administrative areas.)

Tanner has also conducted a cross-sectional analysis of the pattern of US car ownership in 1970 with particular reference to evidence of saturation in high-income areas. Not surprisingly, on the basis of evidence discussed earlier, there was "little evidence of saturation: as income rises, so does car ownership up to the highest levels found, 0.51 to 0.54". This situation was found to apply in three different types of residential area.

Following up this line of enquiry, on the basis of a saturation level in licensed drivers, Tanner noted no evidence of a reduction in the growth rate of US licences. The levels for 1970 still remained well below the level of 90 per cent of all persons aged 17–64 years and 50 per cent of those aged 65 and over. It was noted that, if these rates were achieved by the year 2010, then the overall car ownership level would be 0.63 vehicles per person.

Aggregate Causal Models

Developments in aggregate econometric procedures have mainly involved the derivation of mathematical formulae using national or area cross-sectional data on incomes and other independent variables in place of "time". As in Britain, two common problems have been assumptions on temporal stability of empirical relationships and the effect of multicolinearity of variables.

A number of early models introduced additional variables (to income). Lansing and Mueller (1964) produced a cross-tabulation model, using time series data from US cities and distance from Central Business District as a proxy for level of public transport service. Kain (1964) and Leathers (1967) added several socio-economic zonal characteristics as independent variables, namely present level of car ownership, house type, mode of travel to work, and distance of residence from work. Both these studies were concerned with the "entire mobility sequence" rather than car ownership in isolation.

Beesley and Kain (1964) successfully applied regression analysis to explain car ownership per capita in a sample of US cities (in 1960) in terms of residential density and median family income, claiming that there was no objection to the application of the equation in British cities.

On a wider, national basis, Bottiny (1966) examined trends in car ownership over time in relation to income and driver licence changes with particular reference to evidence of saturation. Noting the earlier comments on saturation by Schmidt and Campbell (1956) and Kanwit, et al. (1960), Bottiny extended the examination to a number of states exhibiting contrasting geographical and economic characteristics. The analysis was mainly concerned with explanation of past trends rather than prediction. Observed differences in state growth rates were explained in terms of variations in income growth, residential density, level of urbanisation and existing level of availability of autos. In several states a trend towards saturation was inferred. At a county level, plateaus were noted in several instances, though all could be explained by special characteristics. Bottiny's study makes an interesting comparison with the more recent work of Divey (1979), which has exposed clear evidence of convergence towards an implied saturation level in discrete areas in the USA.

The complexity of economic models tended to increase in order to exhaust the supply of hypothesised explanatory variables. Dyckman's aggregate-demand model for automobiles (1966) attempted to pull together the earlier work of economists noted above using more up-to-date national data (1929–1962). Using multiple regression analysis demand was explained in terms of income, new-car sales, auto-stocks and credit terms. For the period 1959–1962 these four terms explained 94 per cent of the variation. Suits earlier work on price and income elasticities were confirmed (price elasticity was very low) but the credit variable was found to be highly significant. General work by Evans (1969) and Hymans (1970), though not concerned solely with automobiles, provided further examples of consumer price modelling at an aggregate level. So too did Wyckoff (1973), with the development of an aggregate "stock adjustment model". Annual aggregate demand data for the US 1950–1970 was used as a basis for an explanation of the demand for new cars in terms of existing car stock and observed rate of replacement subject to the restraining effect of consumers' income. The model placed heavy emphasis on the user's perception of cost, noting that most potential buyers already own cars and are well aware of personal costs as a result of hire purchase commitments. As with many academic models, however, no future year forecasts were produced.

The period since 1973 has been characterised by the production of a series of complex aggregate models. There has been a heavy emphasis upon forecasting the consumer market for automobiles, with consequences for car producers in terms of the number of vehicles required and design features. Thus, the value of these general aggregate predictions as a basis for transport planning is limited and, as in the UK, disaggregate models are finding greater favour in this field. The authors of most aggregate demand models refer to the need to accept that there will not be a drastic change in the nature of sale or use of cars in the foreseeable future, and that trends in time series data upon which they are based are typical and unlikely to change significantly in the future.

Chase Econometrics Model of Automobile Demand (1974) is a major example of this recent generation of complex econometric procedures. Using national data for the years 1957–1973 two forecasts are produced: total new passenger registrations and market share for each of five subgroups of vehicles ranked according to size (subcompact, compact, intermediate, standard, and luxury). A weighty regression equation gives the first forecast, new car sales, in terms of ten variables. These are (using a 1958 price base):

1) personal disposable income
2) transfer payments to persons
3) consumer price index
4) consumer price index for new cars
5) rate of unemployment
6) a dummy variable to represent auto strikes in 1964, 1967 and 1970
7) stock of passenger cars on a new-car-equivalent basis
8) consumer price index for gas and oil
9) an index of credit rationing
10) a dummy variable for cancellation and reinstatement of investment tax credit

It is estimated that these variables account for 96 per cent of the observed variation in new car sales 1957–1973.

In the second model distribution of market share is explained using one of five regression equations, according to the class in question, in terms of:

1) rate of unemployment
2) consumer price index for gas and oil
3) miles per gallon for particular subclass of car
4) average price of car in subclass
5) average price of all new cars
6) time (a time trend)

Thus, the market share for the intermediate class of car is derived from the following multiple regression equation:

$$PCT_{IM} = -1.56 + \mathbf{0.443}\,\frac{MPG_{IM}}{MPG_{SC}} - 0.70\,\frac{P_{IM}}{P_{ST}} + 1.87\,\frac{PCIGO}{PCI} + 0.031\,T \qquad (6.2)$$

$\bar{R}^2 = 0.89$

where PCT_{IM} = per cent of total new car sales in intermediate class
$PCIGO$ = consumer price index for gas and oil
PCI = consumer price index
MPG_{IM} = miles per gallon for class

MPG_{SC} = miles per gallon for subcompact class
P_{IM} = average price of new car in class
P_{ST} = average price of new car in standard car class
T = time trend

Not surprisingly, this model has been subjected to considerable academic analysis and criticism. Leaving aside the validity of such a rigid model in an age of increased uncertainty, the main technical discussion has surrounded the inevitable problem of correlation between some of the large number of independent variables (e.g. price and fuel consumption).

Similar problems have been encountered with other contemporary aggregated regression models. Kulash (1975) produced a model similar in concept to that described above but with a level of disaggregation in the procedure prior to aggregation for final national projection. Three aspects of the automobile market are covered: automobile sales, market shares for size groups, and scrappage rates.

The sales model is of particular interest because it involves a technique used in many disaggregate models, namely the use of car ownership/household income probability curves. Total "fleet size" in a given year is estimated as a function of the income distribution.

$$O_t = \sum_I H_I P_{It} R_t \tag{6.3}$$

where O_t = target national automobile ownership in year t
H_I = auto ownership per household of income I
P_{It} = fraction of total households having income I in year t
R_t = total number of households in the USA in year t
I = household income, the summation being made across all income groups

H_I was previously estimated (for 1970) using the equation

$$H_I = 0.01786 \, I^{0.4743} \tag{6.4}$$

where I = income in 1967 constant dollars

A more detailed breakdown of the nature of the "fleet" is obtained by estimating demand for new cars (and resultant scrappings) on the basis of the generalised cost of purchase (includes price of new car, price of gasoline, and fuel economy of vehicle). As in the case of the Chase model, several size classes (three in all) are involved and a similar "market share" procedure is cited.

The Ayres (1976) model is similar to that of Chase in concept and method, although fewer classes of vehicle and independent variables are considered. So, too, is the complex Wharton Stock Adjustment Model (1977). This contains five sets of statistically derived relationships to explain the future composition of the automobile market, in terms of:

1) desired stock
2) desired stock by size (class)
3) new registrations
4) new registrations by size (class)
5) scrappings

Although the model contains many equations they all relate to at least one of these five sets. Figure 6·1 is a simple flowchart of the model. The model was calibrated using 1972 cross-sectional data.

A large number of independent variables are considered, including number of licence holders, number of persons aged between 20 and 29 years, percentage of population living in metropolitan areas, and so on. Cost considerations are combined in the form of one variable (cost per mile) containing fixed and variable elements. In addition to new purchases and market share for each size class, the model can be used to produce estimates of used car ownership by size class and thus the effect of existing stock characteristics on consumers' purchase choice.

The Wharton model has attracted criticism particularly related to its aggregated form and the failure to "explain" the pattern of ownership in terms of socio-economic characteristics at the household level. Consequently, as noted by Ben-Akiva *et al.* (1980):

treating heterogeneous vehicles as if they are identical can be quite severe. For example, to characterize the fictitious 'mid-size' car, the Wharton model takes a sales weighted average of vehicle prices within this class. Since sales are endogenous variables, use of

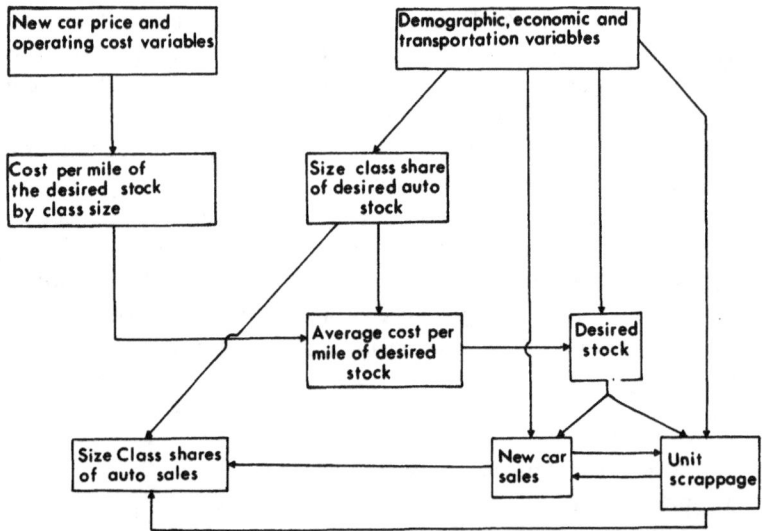

FIGURE 6·1. Simplified flow chart of the Wharton stock adjustment model.

this construct can easily lead to spurious findings. To see this, assume that for exogenous reasons sales of the lower priced mid-size cars increase and sales of the higher priced ones decrease by a like amount. The result will be a fall in the mid-size price index but no change in total mid-size sales at all. From this one might improperly conclude that price does not effect sales volume.

Golomb and Bunch (1979) developed parts of the Wharton model to produce a procedure for specifying future planned and unplanned events in the form of a Future Automobile Population Stochastic Model (FAPS). In similar fashion to the approach used more recently by the UK TRRL, there has been some attempt to allow for variations in future scenarios in the form of a complex computer program. Figure 6·2 shows the structure of a typical "scenario tree" employed in this probabilistic future.

The continued criticism of aggregate modelling, particularly at national level, has been joined by Downs (1979). Citing US Department of Transportation forecasts, it has been possible to expose significant variation in projections which were derived by three different methods. Of particular note is the variation resulting from the applica-

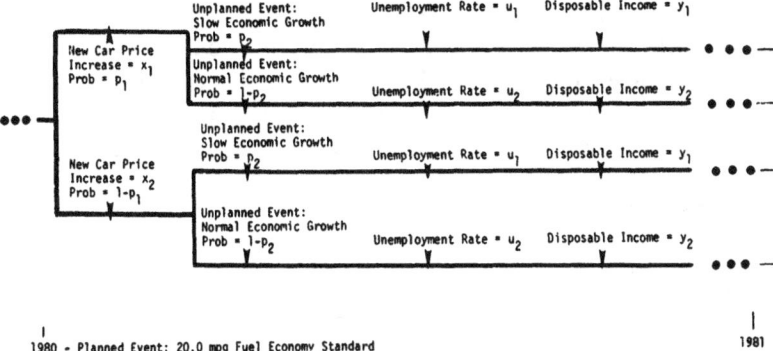

FIGURE 6·2. Example of a scenario tree. (Source: Golomb and Bunch, 1979, p. 49.)

tion of the multiple regression technique to the same problem but using data sets based on different time ranges.

Disaggregate Modelling

Most disaggregate models use the household (a behavioural "car-purchasing" unit) as a source of base data. In most cases projections are aggregated to zonal level for input into urban or regional land-use transportation studies. As with aggregate modelling, early work by economists has been more recently extended by a wide range of academic, commercial and governmental workers.

An early disaggregate approach, previously noted, was used by Farrell (1954), and was further developed by Janosi (1959), Huang (1964) and Lansing and Hendricks (1967). However, in the past decade, the popularity of this concept has increased with a high degree of experimentation in terms of size and geographical characteristics of data base, choice of variables, and level of disaggregation.

Eggert and McCracken (1966) in their summary of US econometric methodology made it clear that long term trends were significantly related to personal incomes and the size of the car population (in relation to human population). But it was acknowledged that other factors could have some effect, namely household structure, residential location and level of education. In line with the increasing application of disaggregate forecasting in regional transport studies the logical development was the introduction of local and household economic models. Bennett (1967) carried out regression analysis on cross-sectional US consumer finance data for 1955–1957, disaggregating down to household level and regarding households as "spending units". A large number of explanatory variables were tested, with the following proving significant:

I	Income	
NA	Number of adults in household	positive
RL	Rural location	
Ag	Age (65 and over)	
NC	Number of children	negative
ML	Metropolitan location	
NW	Proportion of non-whites	

For 1955, the following relationship emerged:

$$C \text{ (consumption)} = 0.07I - 0.121Ag + 13.6NA - 13.9NC$$
$$-105ML + 56.5RL - 113NW \qquad (6.5)$$

This theme of the household as a "spending unit" was accorded even more complex treatment by Cragg and Uhler (1970). Using cross-sectional US data for 1961 and 1962 a multiple regression model employed the following list of variables to explain household demand for automobiles:

1) Disposable income in year preceding survey
2) Disposable income in year preceding
3) Square of disposable income in most recent year
4) US savings bonds in 1960
5) Savings accounts in 1960 } combined as "total liquid assets"
6) Checking accounts in 1960
7) Remaining total instalment debt
8) Age of head of spending unit
9) Number of adults in spending unit
10) Number of children in spending unit
11) Dummy for residence in central city

12) Dummy for residence more than four miles from central city
13) Dummy for holding one or more cars at previous survey
14) Dummy for holding two or more cars at previous survey
15) Value of "first" car at previous survey
16) Value of additional cars at previous survey
17) Age of "first" car at previous survey
18) Age of "second" car at previous survey

Clearly, the use of such a model for long-range predictive purposes is almost impossible. Not only are exogenous predictions concerning most of the "independent" variables liable to error, but there is likely to be some degree of autocorrelation between them. The analysis omitted many of the factors which, by 1970, were known to influence household car ownership rates and which were subject to additional future uncertainty. Some of these had been cited earlier by Eggert and McCracken as "the possible revival of rapid transit in metropolitan areas, the rapidity and direction of the shift of people between suburbs and central city, the current decline in birth rates, the sharp upward trend in leasing, including daily rentals, the further adaptation of cars to meet specialised consumer markets, and the continued improvement of roads, traffic control systems, and parking facilities."

A number of other workers contributed to the debate during the early 1970's, either in the form of specific car ownership models (Hoxie, 1970) or wide travel demand studies (Mundarri, 1972). Dunphy (1973) conducted a specific study of the influence of transit accessibility on ownership patterns. Using 1968 data from the Washington D.C. region she demonstrated a significant relationship between automobile ownership of different households and the transit accessibility to employment of those households. Empirical evidence (from an independent bus passenger survey in Reston, Virginia) indicated that a number of riders had reduced the number of cars owned by their households as a result of improved services. This study parallels a continuing discussion in the USA and elsewhere into the effects of car ownership on transit usage but was unusual in its advocacy of large scale investment in public transport facilities as a dynamic influence on levels of auto ownership and use. The work was purely analytical; no forecasts were produced.

Burns *et al.* (1975, 1976) have also examined the influence of changes in transit availability as well as other spatial activity patterns. This resulted in the formulation of a complex multinomial logit model defining the probability of homogeneous groups of households choosing to own a specific number of automobiles. A large number of variables were considered, including the relative attractiveness of transit services. Recent work by Golob and Burns (1978) using Detroit data has produced some conclusions on the relative importance of transit services, in statistical form. Analysis of cross-sectional data for 1965 revealed that elasticity of automobile ownership with respect to automobile travel times was 0.1 (inferring that a 10 per cent decrease in travel times for automobiles would cause a one per cent increase in car ownership). The same data set revealed an elasticity of automobile ownership against public transit travel times of 0.05 (one half that of automobile travel times) and a corresponding elasticity of ownership with respect to income of 0.3 (three times that of automobile travel times). Thus, the authors claimed to have shown that policies aimed at speeding up traffic on the one hand or public transit on the other could have an effect on overall car ownership levels.

Disaggregate Probability Models

Other recent research in disaggregated modelling has tended to be concentrated on the development of logit probability functions, usually including a large array of household characteristics as explanatory variables. Such models have been produced by Lerman and Ben-Akiva (1975), Kain and Fauth (1977), Johnson (1978) and Train (1979).

Lerman and Ben-Akiva have been concerned with the place of auto ownership in the overall accessibility cycle. Noting that the relationship between automobile ownership and traditional transport forecasting had usually taken the form shown in Figure 6·3 their work involved a call for a change in direction in this field of research. No longer, they said, should it be relegated to a side calculation with single models relying on trend extrapolations or correlations in place of a strong causal theory.

Few studies have ever addressed the basic behavioural factors underlying the household's automobile ownership decisions and still fewer have attempted to embody a be-

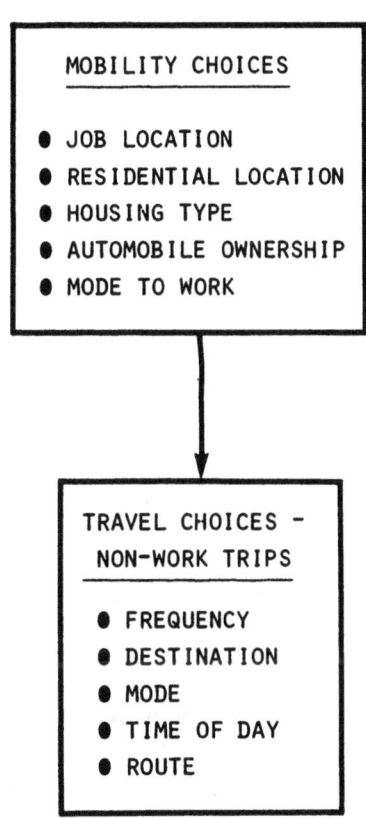

FIGURE 6·3. A conventional car ownership/travel sequence. (Source: Ben-Akiva and Lerman, 1976.)

havioural perspective in a valid econometric model. Furthermore, no modelling efforts have been attempted to incorporate the interaction between the way in which households choose the number of automobiles they wish to own and the other transport-related decisions they make in a behaviourally consistent way.

The response was to develop a series of models "which represent automobile ownership behaviour in a way which is consistent with a credible theory of the choice process at the household level". This involved the study team attempting to integrate the car purchase decision into the overall framework of household decision making, taking

108 CAR OWNERSHIP FORECASTING

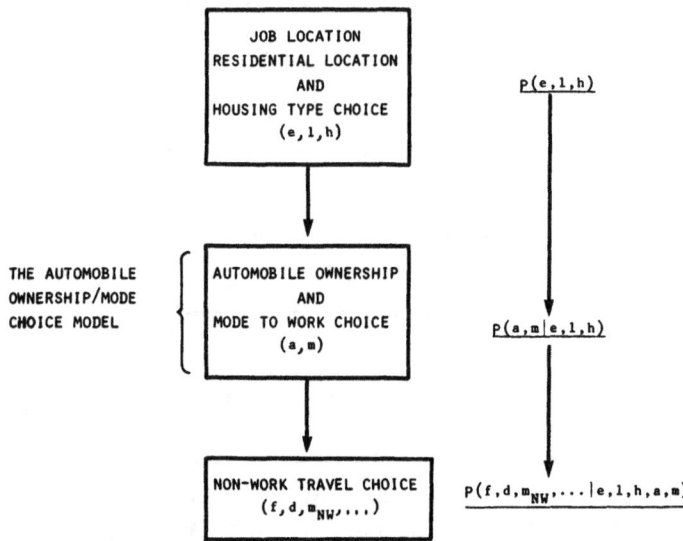

FIGURE 6·4. The Ben-Akiva/Lerman joint choice model. (Source: Ben-Akiva and Lerman, 1976.)

into consideration the long term indirect perceptive environment of the household unit, as well as the more obvious and well-tried variables which were the common base of many earlier models. The result was the application of a multiple choice procedure, the multinomial logit model, in which car ownership is one of a number of choices contained in a wide choice hierarchy itself influenced by causal socio-economic variables, (see Figure 6·4). Using a Washington Metropolitan Area (1968) home interview study as the data base, the following model form was produced to describe the joint auto ownership/mode/choice decision:

$$P(a, m) = \frac{\exp V_{am}}{\sum_j \sum_k \exp V_{jk}} \qquad (6.6)$$

where V_{am} = utility of owning x autos and using mode m for trip to work

a and j = alternative auto ownership levels

m and k = alternative modes, and
$P(a, m)$ is the probability $P(a, m/e, 1, h)$.

Unlike certain models which rely on zonal averages at base year calculation stage, this procedure is truly disaggregate to the household level. Of course its degree of validity is based on the assumption that observed probabilities will remain stable over time and place, and that uncertainties in the rate of change in key variables can be accommodated. This particular problem is catered for in the form of the introduction of mathematical elasticities designed to reflect various future transportation and economic "policy scenarios". A summary of the results of application of alternative "transit scenarios" is shown in Table 6.1 below, with particular reference to the impact upon expected car ownership per household.

The logit probability procedure has been further developed by Train (1977, 1979). In contrast to the approach of Lerman and Ben-Akiva, this involved a specific study of car ownership in isolation from the overall travel picture. The basic form of the model attempted to summarise ownership-choice probabilities as:

$$P_i = \frac{\exp \beta X_i}{\sum_j \exp \beta X_j} \qquad (6.7)$$

TABLE 6.1

Policy Scenario	Description	Expected Auto Ownership	% Change in Auto Ownership From Base Case
Base Case	No transit available	1.825	—
Case 1	Low level transit	1.710	6.3%
Case 2	Good service (e.g., rail rapid transit) for work trips, Case 1 transit service for shopping trips	1.660	9.0%
Case 3	Extremely high quality transit service (e.g., dense areawide personal rapid transit)	1.622	11.1%

(Source: Ben-Akiva and Lerman, 1976, p. 1–15.)

where P_i = probability of choosing to own i automobiles (the number can be 0, 1, 2 or 3 +)
X_i = vector of explanatory variables which affect the choice of owning i variables
β = vector of parameters, which is estimated.

A wide range of explanatory variables were used (income, accessibility, etc.) but none were totally new to the field in terms of general concept.

The same form of model was employed by Lave and Train (1978) as an automobile choice procedure. Probability of choice between ten classes of auto (classified by size and weight) was explained in terms of a similar large range of explanatory variables. Use of both Train models is limited by problems of prediction of change in independent variables as well as likelihood of auto-correlation between them. The authors themselves appreciate some additional problems, resulting from an attempt to quantify variables related to personal taste and perceived value (e.g., "feel", prestigious appearance). Nevertheless, the introduction of these problems into what has been, until recently, a field dominated by normative statistics, gives an encouraging pointer to future direction of research.

Johnson (1978) uses a similar form of model to explain household choice of age and number of automobiles. Two cross-sectional national surveys of consumer expenditure (1959 and 1970) were analysed to ascertain how well components of demand (new and used cars) can be explained by socio-economic variables including family size, age of car and age of head of household. Maxmimum likelihood estimates were derived for the pattern of overall car ownership rates and "automobile age choices" (secondhand car ages). Increases in household income were found to effect positively a probability of multiple ownership as well as the probability of owning a newer car. As age of head of household increases the probability of car ownership is reduced. Family size is positively correlated with car age and overall levels of ownership.

Refinement of disaggregate models continues. Manski and Sherman (1980) have sought to improve Train's logit probability model. In response to recent US legislation on fuel economy standards for new vehicles Ben-Akiva, Manski and Sherman (1980) have sought to introduce new policy assumptions into existing disaggregate behavioural models. Although some change in the characteristics of vehicles is expected, the overall number of vehicles in use is not expected to deviate significantly from earlier projections.

Despite the growing acceptance of the superiority of disaggregate models, there remain a number of limitations related to their validity and use. Most procedures are normative in that they assume a free flow of information across a population which is omniscient with regard to all possible economic, social and physical choices available. Many statistical transportation models falsely assume populations to be composed of perfect *Economic Men*. It has been demonstrated that in many areas where individuals or households enjoy some degree of choice (e.g. choice of mode, choice of route on a network) they are often unaware of all possible options and, even in an environment of perfect information, perception of cost may vary considerably from reality. Variation in human perception can be measured in temporal, spatial and socio-economic terms. The problem can be considerable in the field of vehicle ownership partly because of the wider non-economic role played by car-ownership in society as well as the self-fulfilling dual effect of a concerted advertising system and an established pattern of heavy investment in a highway system designed specifically to favour (or cater for) private auto use (the planning loop syndrome).

Nevertheless, several observers have taken account of present limitations and suggested future directions for research. Train (1979) has suggested that more realistic measures of consumer behaviour could be achieved by conjoint analysis, to approach a sample of potential buyers with a number of possible choices and to rank preferences. This is initially intended to develop a class choice model but could be extended to the general decision to buy a first or replacement automobile bearing in mind the continuing problem of quantification of non-numerical variables. Ben-Akiva *et al.* (1980) have re-emphasised the need to strengthen the interconnection between ownership and pattern of use with particular reference to future development in fuel efficiency.

Recent National Projections: Some Underlying Trends

At the national level the federal Department of Transportation has produced regular forecasts of car traffic over a number of years (e.g. *US Department of Transportation, National Traffic Forecasts*, 1974). These forecasts are intended to be advisory; US states and metropolitan authorities enjoy more autonomy than British counties in this respect. Vehicle ownership is regarded as a major variable influencing overall demand for road travel, and some specific assumptions on future levels have been drawn. Two aspects have been central to the discussion: saturation levels and the pattern of multi-car ownership. The growth in single car ownership has shown signs of levelling over the past two decades while the tendency towards second and third cars per household has become more noticeable. However, even in these categories, the rate is expected to decrease in line with a reduction in the ratio of driving age persons to vehicles. In 1940 the ratio was 3.06, by 1972 it was 1.28 and the projection for 1990 (1974 based) is 1.12.

It should be noted, of course, that the present level of car ownership in the United States is already well in excess of the saturation level assumed in several European models, including the UK national procedure (in 1976 the US rate was 0.51 cars per person). In several states the figure already exceeds 0.55 yet most US projections still imply considerable growth. The national forecast for licensed drivers (which in 1975 stood at 126.5 million) is 161.2 million for 1990. This represents a growth in the overall percentage of population licensed to drive from 80.3 per cent in 1975 to 85 per cent by 1990.

Perhaps of greater significance is the overall distribution of vehicles to households. In spite of a consistent growth in one-car and multi-car ownership in recent years a considerable proportion of US households still have no access to private vehicles (see Figure 6·5). Even allowing for physical limitations in households consisting entirely of aged persons, it must be assumed that a proportion of the national population would purchase first cars if they were sufficiently affluent. Alternatively, the improved mobility resulting from ownership may be perceived as a justification for a redistribution of limited household expenditure. Figure 6.5 indicates the main category of growth in recent decades has been in multi-car ownership on the part of the more affluent section of the community.

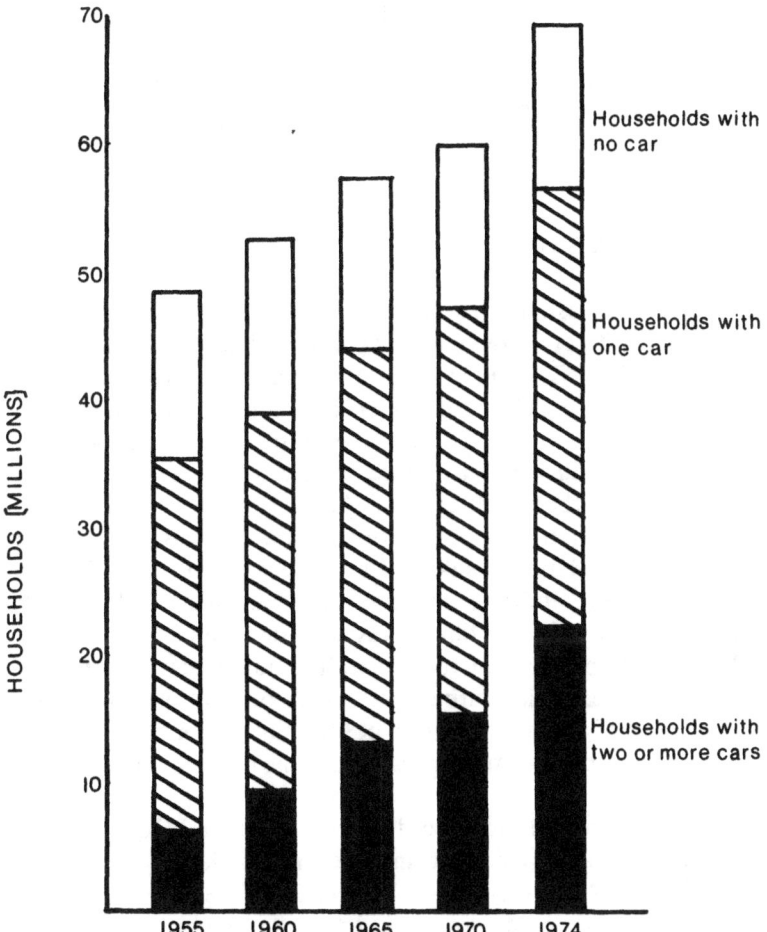

FIGURE 6·5. Changes in the pattern of US household car ownership 1955–1974.

Any work in other developed countries which cites the USA as an example, particularly in a discussion of saturation levels, should acknowledge that different socio-economic and political conditions may prevail. A saturation level in the USA, evident by a stable person/car ratio, is likely to conceal considerable suppressed demand on the part of certain socio-economic groups. If saturation is reached in the UK, and assuming its narrow income distribution does not change significantly, then this is likely to reflect different demand conditions. There is less likelihood that a large section of the community will be experiencing economic frustration in a situation of perceived "stability" in car ownership.

Forecasting in Australia

The use of forecasting methods in Australia provides an interesting comparison with other developed countries. There are similarities to Canada and New Zealand in terms of the variables likely to influence car ownership: a high level of affluence, lower overall population density counterbalanced by a heavy concentration of population in a small number of urban areas, these being characterised by expansive low density suburban development. An earlier pre-war phase of suburban growth (based on railway corridors) was followed by superimposition of auto-based residential and industrial development on a concentric basis, infilling earlier green "wedges" and extending beyond the limits of the initial railway commuter zone.

In 1971, 71 per cent of the Australian population resided in the seventeen largest cities (urban areas with populations in excess of 40,000). Out of a total population of fourteen millions, more than six millions live within commuting distance of the two cities of Sydney and Melbourne.

Despite a recent trend towards increased Federal Government involvement in road transport, trunk roads remain largely the responsibility of the six Australian states. Projection of inter-urban road traffic has been based on crude growth factor methods which have not contained any particular allowance for variation in vehicle ownership beyond an inherent acceptance that past trends are likely to continue.

In contrast to the simple methods employed at state and federal level is the pattern of prediction employed in the larger urban areas. Most of the large cities, particularly the state capitals (and Canberra), are autonomous to a much greater extent than their European counterparts. Partly as a result of this, they have all been subjected to standard land use/transportation studies containing a variety of car ownership prediction techniques.

Urban land use transportation studies commenced in 1961 with work on Canberra and culminated with the publication of a massive report on Sydney in 1974. In the intervening period, more than twenty cities have been covered by a number of agencies employing a variety of methods. In some smaller centres, individual car ownership projections were not provided, the emphasis being on "traffic growth" between land use zones.

Predictably, most projections were based on the recently criticised methodology in general acceptance during the 1960's, namely an area-wide total achieved by means of an aggregated extrapolation, coupled with some attempted zonal estimation employing either the same technique or a crude form of explanatory model. In one of the earlier studies (Brisbane, 1965) no area control was used, zonal projections being based on a standard multiple regression model employing population, income and residential density as explanatory variables. In other early studies, namely Canberra (1963), Hobart (1964), Toowoomba (1966) and Adelaide (1968) a simple area-wide extrapolation was employed as a means of producing an overall control figure. Later, a logistic curve was preferred, notably in Launceston (1968) and Hobart (1970), at both area and zonal level. In the Perth study (1970) a hyperbolic curve was employed at area and zonal level, whereas the study team in Bendigo (1972) used a Gompertz curve for similar purposes.

The Townsville study (1966) had employed Tanner's UK logistic curve at area level with zonal values being based on extrapolation from an empirical curve based on income and zone location. This method was also used to produce area controls in the two largest studies conducted, in Melbourne and Sydney. The Melbourne project (1964) involved an empirical base year study (1964) which indicated an apparent saturation situation in certain high income zones. The average car ownership level in such zones, 0.38 cars per person, was taken as a likely saturation level with a design year (1985) projection of 0.354. Zonal projections also employed a similar logistic curve with controlled modifications to allow for variations in income, employment characteristics and level of public transport services.

The widespread use of extrapolatory curves in Australian studies exposes the main weakness of this method, namely the assumptions on and effect of saturation level. Even allowing for variation in social and geographic characteristics between individual cities, it is unlikely that the Melbourne saturation level would eventually prove to be 0.38 cars per person while the equivalent in Perth was 0.65 and Bendigo 0.74 (as indicated by their respective land use/transportation plans).

The most extensive urban land use transportation study was conducted in Sydney in the early 1970's (Sydney Area Transportation Study, 1974). Prior to this, Brain (1970) had produced forecasts for New South Wales (NSW) using the TRRL logistic curve method. Using historical data for NSW for the period 1950–1968 and a saturation level of 0.55 (based on California) the following equation was derived:

$$\text{Vehicles per person} = \frac{0.55}{1 + 0.98198\, e^{-0.0855 t}} \quad (6.8)$$

where t = year – 1968 (the value for 1975 would be 1975 – 1968 = 7)

Brain's work formed the basis for controlling regional and local models and forecasts in the Sydney study. As was usual there was extensive analysis of the effect of "causal" variables including work force participation, residential density, income and accessibility to public transport. A "best" regression equation was derived to explain zonal variations using two variables, residential density and family income. Because of difficulties in estimating future growth in incomes and residential density it was decided to base zonal car ownership projections primarily on Brain's logistic curve, it being assumed that "values of vehicles per person for individual traffic zones were distributed normally around the regional average, which lies on the logistic curve". Zonal estimates were subjected to several "manual adjustments" to allow for variations in residential density and to reduce inconsistencies noticeable when ownership rates for the year 2000 exceeded the control saturation level (see Figure 6·6). Thus, a logistic curve forecast of 0.517 was reduced to 0.493 after adjustment. By the year 2000, it was estimated that 91 per cent of households would own cars and, more significantly, 48 per cent would own at least two cars. Estimates of vehicle mileage for 2000 fell within a range which had been extrapolated previously by the Commonwealth Bureau of Roads (1971).

The Sydney study (SATS) was a late example of a conventional reg-

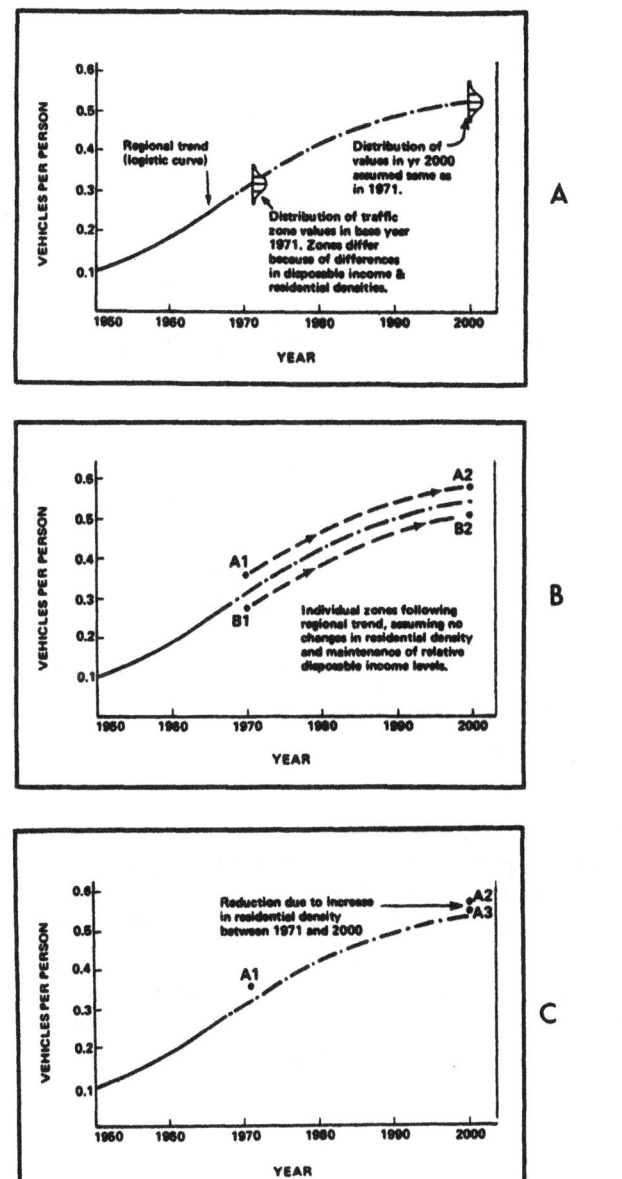

FIGURE 6·6. Logistic curves of car ownership used in the Sydney Area Transportation Study. A: Basic curve. B: Adaptation to allow for zonal distributions. C: Adaptation to allow for variations in residential density. (Based on Figure 3.9, 3.10, and 3.11, *Sydney Area Transportation Study*, Volume II, 1974.)

ional transportation study and is notable for the rigidity of the approach to car ownership, using an extrapolation procedure in preference to regression or stochastic methods. Forecasts of future traffic, and the programme of roads required for its accommodation, have been the subject of lengthy technical and political argument.

As in other developed countries, there has been recent discussion on and application of behavioural models in Australia. Hutchinson (1979) describes the application of an individual household choice model in Adelaide and Canberra. The basic model, using cross-sectional data collected in the two cities, takes the form:

$$P \text{ (Choice)} = \frac{e^u}{1 + e^u} \tag{6.9}$$

where P(Choice) = probability of upper level of choice
 u = some indirect relative utility function of ownership choice. u is formulated as a linear function of parameters representing relevant household characteristics.

The choice structure considered was (on a household basis),

1) Car ownership/non-ownership
2) One car/multi-car
3) Two cars/three or more cars.

The result of a lengthy analysis involving selection of parameters and calculation of coefficients has been a questioning of the importance of income as the highly significant variable used in several UK models, with new emphasis upon household structure. Bearing in mind the high levels of real income and existing car ownership in the cities in question (compared to the UK), this corresponds with the trend experienced in the USA, which suggests that the determining effect of household income is reduced as saturation is approached.

REFERENCES

Ayres, R., et al., Automobile Forecasting Models, International Research and Technology Corporation Working Paper IRT-446-R. Washington DC, 1976.

Bayley, J., Transportation Plan — Melbourne 1985, *Traffic Quarterly*, January 1971, pp. 73-85.

Beesley, M.E., and Kain, J.F., Urban form, car ownership and public policy: an appraisal of 'Traffic in Towns', *Urban Studies*, I, No. 2, 1964, pp. 174-203.

Ben-Akiva, M.E., and Lerman, S.R., *A Behavioral Analysis of Automobile Ownership and Mode of Travel, Volume III*, US Department of Transportation, 1976.

Ben-Akiva, M.E., Manksi, C.F., and Sherman, L., A behavioral approach to modelling household motor vehicle ownership and applications to aggregate policy analysis, Paper presented at the World Conference on Transport Research, London, April, 1980.

Bennett, W.B., Cross-sectional studies of the consumption of automobiles in the United States, *American Economic Review*, LVII, 1967, pp. 841-850.

Black, J., A review of the technical aspects of Australian urban land use/transportation studies, reprinted in Webb, G.R., and McMaster, J.C., *Australian Transport Economics*, ANZ Book Company, Sydney, 1975, pp. 307-331.

Bottiny, W.H., Trends in automobile ownership and indicators of saturation, *Highway Research Record*, No. 106, 1966, pp. 1-21.

Brain, R., A forecast of car ownership in New South Wales, *Australian Planning Institute Journal*, VII, 1970, p. 22.

Burns, L.D., Golob, T.F., and Nicolaidis, G.C., A theory of urban households automobile ownership decisions, *Transportation Research Record*, No. 569, 1975, pp. 56-74.

Burns, L.D., and Golob, T.F., The role of accessibility in basic transportation choice behaviour, *Transportation*, V, 1976, pp. 175-198.

Chase Econometrics Associates, *The Effect of Tax and Regulatory Alternatives on Car Sales and Gasoline Consumption*, NTIS Report PB-234622, 1974.

Cragg, J.G., and Uhler, R.S., The demand for automobiles, *The Canadian Journal of Economics*, III, August 1970, p. 386.

Deutschman, H.D., Auto ownership revisited: a review of methods used in estimating and distributing auto-ownership, *Highway Research Record*, No. 205, 1967, pp. 31-49.

Divey, S.T., Regional and national convergence to common car ownership levels, *Transport and Road Research Laboratory Supplementary Report SR 463*, Transport and Road Research Laboratory, Crowthorne, 1979.

Downs, A., The automotive population explosion, *Traffic Quarterly*, XXXIII, July 1979, pp. 347-362.

Dunphy, R.T., Transit accessibility as a determinant of automobile ownership, *Highway Research Record*, No. 472, 1973, pp. 63-71.

Dyckman, T., An aggregate demand model for automobiles, *Journal of Business*, XXXVIII, July 1966, pp. 252-265.

Eggert, R.J., and McCracken, P.W., Forecasting the automobile market, in Butler, W.F., and Kuvesh, R.A., *How Business Economists Forecast*, Prentice-Hall,

New Jersey, 1966, pp. 313–333.
Evans, M., *Microeconomic Forecasting*, Harper and Row, New York, 1969.
Farrell, M.J., The demand for motor cars in the United States, *Journal of the Royal Statistical Society* (A), 1954, pp. 171–193.
Golob, T.F., and Burns, L.D., Effects of levels of transportation service on automobile ownership in a case study urban area, *Transportation Research Record*, No. 673, 1978, pp. 137–145.
Golomb, D.H., and Bunch, H.M., Stochastic analysis of future vehicle populations, Highway Safety Research Institute Report No. DOT–TSC–NHTSA–79–20, The University of Michigan, Ann Arbor, Michigan, May 1979.
Hoxie, P., Auto Ownership and the Long-Run Transportation Decision, Unpublished S.M. Thesis, Massachusetts Institute of Technology, Cambridge, Massachusetts, 1970.
Huang, D.S., A microanalytical model of automobile purchase, Research Monograph No. 29, Bureau of Business Research, Graduate School of Business, University of Texas at Austin, 1964.
Hutchinson, M.J., Multi-variable models of car ownership, *Traffic Engineering & Control*, 20, 1979, pp. 399–403.
Hymans, S., Consumer durable spending: explanation and prediction, *Brookings Papers on Economic Activity*, No. 2, Washington DC, 1970.
Janosi, P.E. de, Factors influencing the demand for new automobiles, *Journal of Marketing*, XXIII, April 1959.
Johnson, T., A cross-section analysis of the demand for new and used automobiles in the United States, *Economic Inquiry*, XVI, 1978, pp. 531–548.
Kain, J.F., A contribution to the urban transportation debate: an econometric model of urban residential and travel behaviour, *Review of Economics and Statistics*, XLVII, 1964, pp. 55–64.
Kain, J.F., and Fauth, G.R., *Forecasting Auto Ownership and Mode of Choice for U.S. Metropolitan Areas*, Harvard University, Cambridge, Massachusetts, 1977.
Kanwit, E.L., Steele, C.A., and Todd, T.R., Need we fail in forecasting?, *Highway Research Board Bulletin* 257, 1960, p. 22.
Kulash, D.J., Forecasting long run automobile demand, *Transportation Research Board Special Report No. 169*, 1975, pp. 14–19.
Lansing, J.B., and Mueller, E., *Residential Location and Urban Mobility*, Survey Research Center, Institute for Social Research, University of Michigan, Ann Arbor, Michigan, 1964.
Lansing, J.B., and Hendricks, G., *Automobile Ownership and Residential Density*, Survey Research Center, Institute for Social Research, University of Michigan, Ann Arbor, Michigan, 1967.
Lave, C., and Train, K., A disaggregate model of auto-type choice, *Transportation Research*, XIIIA, 1978, pp. 1–9.
Leathers, N.T., Residential location and mode of transportation to work: a model of choice, *Transportation Research*, I, 1967.
Lerman, S.R., and Ben-Akiva, M.E., A disaggregate behavioral model of automobile ownership, *Transportation Research Record*, No. 569, 1975, pp. 34–55.

Liston, L.L., Projections of Motor Vehicle Registrations, Driver Licenses and Motor-Fuel Consumption to 1990, US Department of Transportation, Washington, DC, 1976.

Manski, C.F., and Sherman, L., Forecasting the fuel efficiency of household motor vehicle holdings, Cambridge Systematics, Inc., Cambridge, Massachusetts, 1980.

Mundarri, D.H., Mode choice for work trips: a key transportation problem, Unpublished PhD dissertation, University of Michigan, 1972.

Schmidt, R.E., and Campbell, M.E., *Highway Traffic Estimation*, Eno Foundation, Saugatuck, 1956.

Sydney Area Transportation Study (SATS), Sydney, NSW, Australia, 1974.

Tanner, J.C., Car ownership trends and forecasts, *Transport and Road Research Laboratory Report LR 799*, Transport and Road Research Laboratory, Crowthorne, 1977.

Train, K., Consumers' responses to fuel-efficient vehicles: a critical review of econometric studies, *Transportation*, VIII, 1979, pp. 237–258.

Wharton Economic Forecasting Associates, *An Analysis of the Automobile Market: Modelling the Long-run Determinants of the Demand for Automobiles*, (3 vols.), Prepared for US Department of Transportation, Transport Studies Center, Cambridge, Massachusetts, 1977.

Whorf, R.P., Models of Automobile Ownership, Ford Motor Company, Detroit, Michigan, 1973.

Wyckoff, F.C., A user cost approach to new automobile purchases, *Review of Economic Studies*, XL, 1973, pp. 377–390.

7 National Theory: Local Practice

PREVIOUS CHAPTERS have outlined the magnitude and range of research activity which has been directed towards the problem of car ownership prediction. This has been matched by a similar effort on the part of transportation planning consultants and other workers at the national, regional and conurbation level. (All notable conurbations and free-standing cities were subjected to conventional transport planning studies during the late 1960's and early 1970's.) It seems appropriate, therefore, to examine the policy of various UK authorities towards the problem of car ownership prediction.

Government involvement in transport planning is conducted on several tiers. At the national level the Department of Transport is the Ministry chiefly responsible for policy development and overall budgetary control in a wide field of activity. National policy is interpreted at the regional level by a Regional Controller of Roads and Transport, whose role is to ensure that county and district authorities follow the correct procedures with regard to transport planning. Trunk roads and motorways are the direct responsibility of central government with new developments being handled by one of several regional Road Construction Units.

Car ownership forecasts are an important input into the technical case for new trunk road and motorway schemes. The Road Constriction Units are primarily responsible for provision of forecasts. During the 1970's, several modelling procedures were used. The earlier National Traffic Model (1975) is now being superseded by regional projections based on the use of the Regional Highway Traffic Model (which contains a car ownership sub-model) employing the 1980 National Traffic Forecasts as an aggregate control on local variations.

In England and Wales, outside London, local transport planning and co-ordination is largely the responsibility of six metropolitan counties and a greater number of "shire" counties. All are highway authorities, having responsibility for classified roads other than trunk routes and motorways. In metropolitan areas public transport services are primarily the responsibility of passenger transport executives

(PTE's) which are under the direct control of the county authorities. District councils have more limited powers in the transport field, though many shire districts operate their own municipal bus services, the viability of which is closely linked to other transport planning issues.

Clearly, all these authorities share an interest in the problem of car ownership forecasting. Even at district level, where there may be no direct need for traffic forecasts, there is a need to consider future trends in ownership in such fields as layout of residential areas, development control in housing, industry, commerce, and, more directly, central area car parking provision. Nowadays, few planning issues do not involve some reference to levels of car ownership, even if this is indirect.

County councils consider transport planning matters on both long term and shorter term bases. Structure plans set out the overall strategy for movement of people and goods in the longer term (usually up to fifteen years beyond the base year). The structure plan enables transport planning to be integrated with land use development. Local plans put the structure plan transport policies into effect, containing detailed proposals for highway improvements and other physical changes involving land use.

Short term planning policies are currently contained in the county council's Transport Policies and Programme (TPP). This is prepared annually and is concerned with questions of resource allocation, value for money and scheduling of projects over a three year period, within the framework of structure plans and local plans. The system was introduced in 1974 and has been modified since then in the light of working experience. A continuing aspect of the philosophy behind this procedure is the requirement that proposed infrastructural (and, sometimes, vehicular) changes should be supported by numerical evidence of the future level of demand for transport services.

The TPP procedure is still a relative newcomer to the transport planning system. Nevertheless, even though a thorough nationwide survey of local authority policy and methods is beyond the scope of this work, an examination of a small sample of "typical" local authorities reveals a notable breakdown in the link between the (mainly) academic activity which has been the first concern of this book and the application of results at the grass-roots level. From a survey of about twenty local authorities (with particular reference to their structure plans) it is clear that three types of approach have been adopted. The first, noted below, is a complete omission of any reference to car ownership prediction, a situation which applies to several outlying rural counties. Secondly, there are some semi-rural counties which appear to be following a policy of blind adherence to the Department of Transport official forecasts (largely, as we have seen, based on the work of the TRRL). At the other end of the spectrum is a group of larger authorities which have had the benefit of an intensive transportation study at some time during the past fifteen years and most of which are at present applying a "mixed bag" of car ownership prediction techniques. This may involve attempts at "up-dating" transport study data or adaptation of Ministry forecasts with allowances for expected effects of, and variations in, population density, household size, and income distribution. Not surprisingly, in view of the large population and extensive resources, the most sophisticated procedures are adopted by the Greater London Council where, in common with several other large authorities, professional staff are able to develop and test new procedures and hypotheses for the overall benefit of national planning.

Examples of Local Authority Practice

As noted earlier, an examination of county structure plans has exposed a wide variation in policy towards car ownership prediction. At one extreme are Hereford and Worcester County Council in their *Hereford County Structure Plan* (1976) which, in spite of a lengthy and detailed discussion of numerous highway improvement schemes, makes no mention whatsoever of car ownership or use of projections. This is in a county (old Herefordshire) where a particular problem exists (especially in rural districts within commuting distance of Hereford). It is evident, on an empirical basis, that a considerable number of small households with two plus cars are based in "suburban" villages. In contrast, many working class households (either dependent on agriculture or manufacturing/service employment) are without personal transport. Any application of aggregated present day or predicted values could, therefore, provide a distorted picture of the real levels of car ownership and use and great care should be taken in the matter of equity (when considering allocation of resources to highway schemes instead of public transport services). The problem is common to other rural areas where a high aggregated car ownership level presents the same type of distortion.

Several other shire counties have exhibited almost the same level of simplicity in the matter of car ownership. However, the majority appears to have opted for straight acceptance or limited manipulation of Ministry forecasts which have been produced since the late 1960's. An early urban structure plan, report of survey, for the Brighton area (1972) used the national forecasts of Tulpule (1969) "adapted" for local use. The *Worcester Structure Plan Report of Survey* (1973) also used the TRRL model "on a county basis", a method employed in the *Leicester County Structure Plan* (1973). The *Northampton County Structure Plan (Report of Survey)* (1976) includes a section on car ownership in the form of a straightforward summary of *Technical Memorandum H3/75*, with projections up to 1990, and reference to the TRRL Report LR 650. Other counties which are being increasingly influenced as a result of their proximity to London received similar treatment, as in the *Buckingham County Structure Plan* (1976). The *North Yorkshire County Structure Plan* (Report of Survey) (1976) also employs Department of Transport guidelines, on this occasion from the 1976 *Consultation Document*, to produce a county range of cars per head for the period up to 1995. A large programme of highway schemes contained in the *Devon County Structure Plan* (1979) is justified on the basis of traffic flows which were projected on the basis of the 1978 *Interim Memorandum*. To their credit the authors of several of these plans have highlighted the difficulties associated with forecasting at a time of continued academic debate over the validity of established methods.

Despite good variations in socio-economic characteristics across the county and a noted enlightened planning philosophy, *Durham County Structure Plan* (1979) indicates that Durham County Council has also employed H3/75 and *Interim Memorandum* forecasts. County Durham (as designated by 1974 boundary changes) provides an interesting example of the problems resulting from aggregation of data and predictions at county level. The philosophy behind the new boundaries here was particularly concerned with the inclusion of large tracts of uninhabited Pennine upland intended to "balance" a smaller but much more densely populated industrialised sector in the east. Even here there are notable socio-economic contrasts between former coal mining settlements and the more affluent parts of Durham City and Darlington, as well as two "new towns". The problems resulting from the application of an aggregated forecast for car ownership are as great here as anywhere and, in addition, the particular settlement pattern on the coalfield sections means that the use of population density as a negative constraint on ownership prediction does not have the same force as it would in truly urban centres.

Lancashire County Council has been responsible for work on two plans, the *North East Lancashire Structure Plan* (1980) and the *Central and North Lancashire Structure Plan* (in preparation). Although by no means unique, this authority has used techniques which are basically different from those mentioned so far with the application of a household-based disaggregated category analysis trip generation model (using TPA CATAN computer program). Although there are detailed differences, the general principles of the CATAN model are similar to those used in early US studies and developed by Wootton and Pick (1967). Households in traffic zones are cross-classified on the basis of three characteristics which are thought to influence trip-making behaviour, namely household income, car ownership and household structure. In this version of the CATAN model it was possible to identify 162 different types of household. A car ownership sub-model employed car ownership/income probability curves (Figure 3·3) to produce zonal estimates of car ownership for the base year and future years. Estimates of base year income were derived from a separate procedure which, in the absence of survey data, estimated household income according to socio-economic group of head of household (see Figure 7·1). Variations in future year levels of car ownership were produced to allow for different assumptions on growth in real income over the plan period (ranging from 1.5 per cent per annum to 3.5 per cent per annum). The resultant zonal projections were aggregated to district level, but even in this case the totals were adjusted so as to be compatible with the national aggregate forecasts contained in the *Interim Memorandum* (1978). However, the ratio of variation between zones still remained.

As noted earlier, several urban areas and free standing towns have been subjected to extensive land use transportation plans. Most of these studies involved some form of car ownership modelling, and many form the basis of later structure plans. A typical example is the County of Tyne and Wear, the subject of *Tyne Wear Plan: A Transport Plan for the 1980's* (1972).

As might be expected, in view of the comprehensive nature of the study and its relatively recent publication data, the authors of the *Tyne and Wear Structure Plan* (1976) made full use of its findings with only a limited attempt to update the findings and predictions of the 1972 plan. Despite an established pattern for the car ownership rate in Tyne and Wear to fall consistently below the national average (for reasons of population density and socio-economic variables related to it), and

NATIONAL THEORY: LOCAL PRACTICE 127

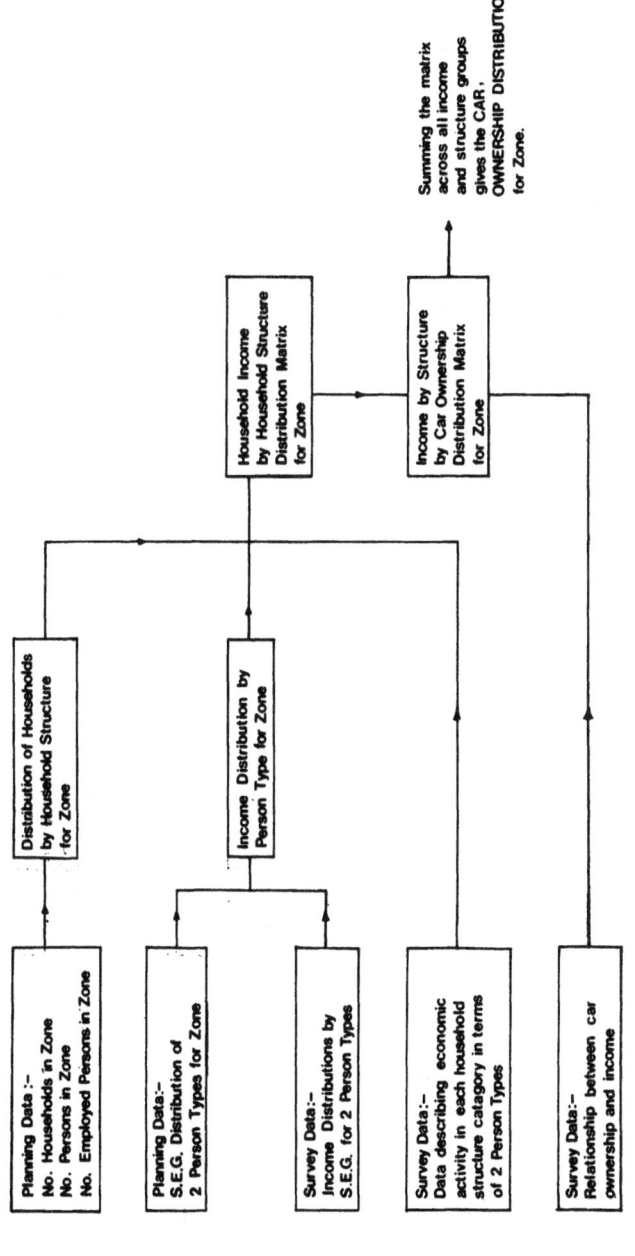

FIGURE 7·1. Stages in the car ownership modelling process used in Lancashire County Council Structure Plans.

128 CAR OWNERSHIP FORECASTING

even below the average for the Northern Region, the plan concludes that "it is not possible to say whether Tyne and Wear will eventually catch up with the national average, but it is probable that as regional income increases car ownership will also increase." Thus, no actual numerical forecasts were provided. The Tyne Wear Plan (upon which the Structure Plan is partly based) was also noteworthy in terms of the limited attention accorded to the problem of car ownership prediction. In contrast to similar work on other conurbations, the survey team used a TRRL logistic curve approach, with modification of ownership rates in the form of cars per worker, rather than cars per head of population. Figure 7·2 shows that the planners expected ownership rates amongst non-manual workers to eventually reach a saturation level commensurate with the national average (based on modified TRRL predictions, despite the qualifications held by Tanner regarding areas — like Tyne and Wear — of higher population density).

Although less important in terms of population and standard metropolitan characteristics, Teesside was also subject to an intensive

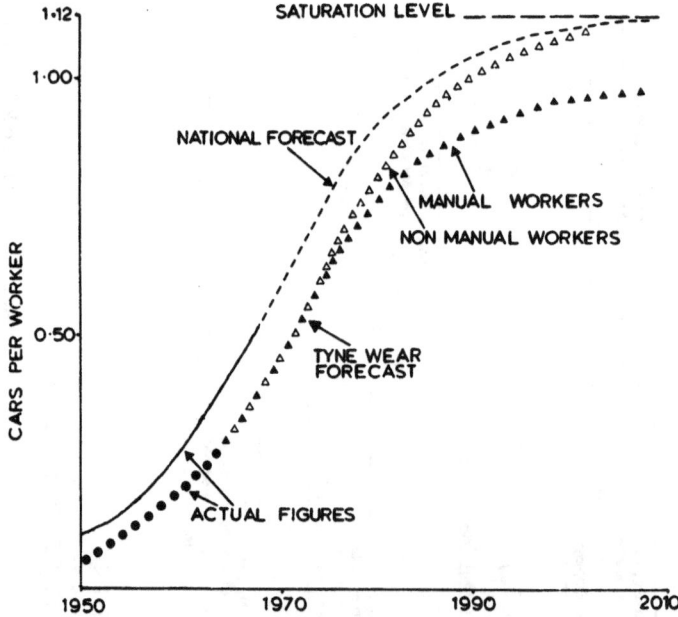

FIGURE 7·2. Logistic curves used in the Tyne and Wear Structure Plan. (Source: *Tyne and Wear Plan*, 1972.)

transportation study, the *Teesside Survey and Plan* (1971). Coupled with use of Ministry revisions of forecasts and some limited individual work, the present day Cleveland County Council has drawn heavily upon the initial data base provided by "Teesplan".

The Teesplan survey was noteworthy in that three separate car ownership prediction methods were employed, the final forecasts resulting from a combination of two of these. The three analyses of future levels of car ownership were:

A) a study of the relationship between household income and household car ownership
B) a logistic prediction based on the TRRL method
C) a regression analysis of ownership, residential density and family income.

Method A involved the use of cross-sectional survey data to derive a linear relationship between income and ownership (see Figure 7·3). In spite of the fact that the relationship between the two variables was considered to be linear (with no curvilinear effect for higher incomes) this method produced "poor" results in the form of a low rate of ownership (0.26 cars per person) for 1991, and was subsequently dropped from the analysis. Method B employed the TRRL logistic curve procedure using the equation:

$$\text{per capita car ownership} = \frac{ax}{x + (a-x)\exp[-(art/(a-x))]} \quad (7.1)$$

where
x = 1966 value of car ownership = 0.128
a = asymptotic value of car ownership = 0.40
r = 1966 rate of growth = 0.09
t = time in years (1966 = 0)

This formula reduced to:

$$\text{per capita car ownership} = \frac{0.0512}{0.128 + 0.272 \exp(-0.1325t)} \quad (7.2)$$

A 1991 value of 0.372 cars per person was produced. There was no further discussion about the justification for choice of the particular saturation level, or the validity of using the 1966 growth rate (a high 9 per cent). Method C involved the derivation of a multiple linear regression equation as follows:

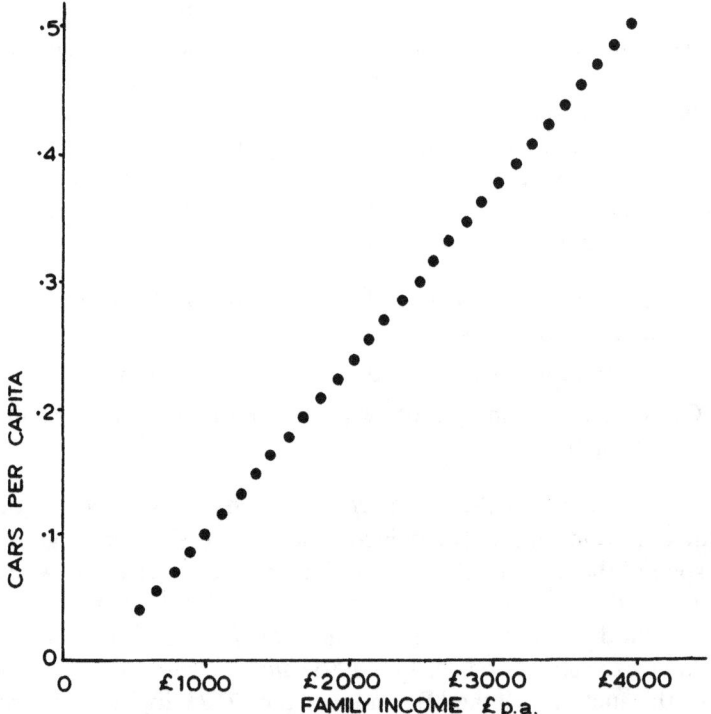

FIGURE 7·3. Car ownership/income relationship used in the Teesside Survey and Plan. (Source: *Teesside Survey and Plan*, Volume 2, p. 270.)

car ownership = 3.78 + 0.83 median family weekly (7.3)
income − 0.10 residential density
(persons per acre).

This produced a car ownership rate of 0.348 for 1991.

Final forecasts for 211 fine zones were produced by a combination of methods B and C. An overall projection for the entire area (0.36) was predicted by using the logistic curve, while zonal values were produced from the regression procedure. This produced zonal rates for 1991 which varied from 0.181 to 0.551. Car ownership rates were then fed into a more complex traffic model as one of eight "land use parameters". In order to accommodate the estimated flows of private and goods traffic a large scale system of new highways was proposed.

The numerical basis of the plan, and the road proposals, were virtu-

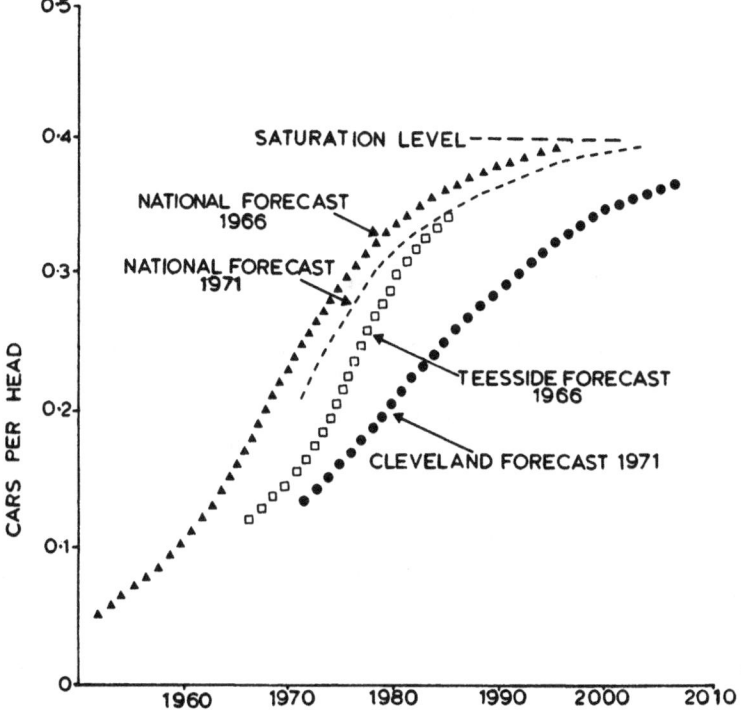

FIGURE 7·4. Car ownership forecasts used by Cleveland County Council.

ally unaltered in their incorporation into the local statutory structure plan (*Teesside Structure Plan*, 1975). Since that date several modifications to county projections have been carried out but, as recently as 1978, these still involved use of the TRRL logistic formula used in the Teesplan study. The effect of a change in the growth rate parameter alongside a retention of the Teesplan saturation level is evident from Figure 7·4.

Studies in Teesside highlight, once again, the problems associated with the application of empirically derived national models disaggregated to county level. "Teesplan" used population density as an independent variable to produce zonal averages and at this level of disaggregation the exercise may have been valid, even allowed for within zone variation. At county level, however, with more aggregation, spatial and social variation is likely to increase. A particular complication arises in Cleveland because the area is unusual in having wide sections of unsettled land given over to industry, dereliction or waste. In addition there is a considerable variation in population density and social conditions. Some of these problems have been recognised and, in common with other authorities, alternative causal models are now taking preference over extrapolation and linear regression. Recent local transport studies have involved use of household category analysis as well as the Department of Transport's household probability model.

New Directions for Local Forecasting

This brief examination of techniques used at county level has exposed a wide variation in methods and philosophy on the part of transport planners. It is clear that there has been a strong tendency to employ national projections, in spite of a continuing academic argument over their value and validity at local level. The problem of local variation in ownership has been a major theme of academic study by Button *et al.*, and has now received renewed Department of Transport involvement, following the publication of the ACTRA Report. In addition to the Department of Transport household model (which is still the subject of development and calibration) Hayfield (1979) has produced further guidelines for workers at regional and county level. This involves a method of forecasting car ownership in local areas using National Travel Survey Data. This has been based on a project which has examined the stability of relationships over time using NTS data for 1965, 1972/73 and 1975/76. It was considered that the NTS was an appropriate base for modelling because it was conducted at regular intervals and included a number of variables allowing relationships to be tested over time.

This method involves a direct application of the Department of Transport causal model, described in Chapter 4. However, allowance is made for different income growth factors for various regions. Although the method represents an attempt at partial disaggregation of a national cross-sectional procedure, its potential is limited by a number of factors, including the need to use fairly large regional units so as to be representative of a range of income groups compatible with the national distribution. So the problem of intrazonal variation, and the degree to which regional averages are truly representative, remains. And, as noted in the context of the 1980 National Traffic Forecasts, locally derived estimates are expected to be compatible with national totals.

An Example of Local Forecasting in the USA

Various methods of local forecasting have been used in the USA. Some have already been described in Chapters 5 and 6. As in the case of transport authorities in the UK, varying levels of sophistication have been applied. Once again, local methods have exhibited a number of common fallacies by way of assumption on the temporal stability of empirically derived models. In common with several other authorities, the State of California Department of Motor Vehicles is currently using a multiple regression technique to forecast future registration figures. Historical data for the period 1960–1977 was used to produce the following equation for passenger vehicles:

$$Y = -5,873,000 + 1.4363X_1 - 1,217.67X_2 \tag{7.4}$$

where Y is the number of passenger vehicles
X_1 is the Californian civilian population aged 16–64
X_2 is the gross national product in billion dollars.

The coefficient of determination (R^2) is 0.9906, illustrating that these two independent variables accounted for more than 99 per cent of the variation in vehicle numbers.

In producing a set of forecasts for the period 1980–2005 independent estimates of gross national product were produced by a simple regression analysis. Thus, the accuracy of both vehicle and national product forecasts is dependent upon assumptions of linearity in the relationships under examination and this, as we have seen, has been a constant source of controversy in the general car ownership debate.

REFERENCES

Department of the Environment, *The National Traffic Model — An Introductory Guide*, DOE, London, 1975.
Department of the Environment, Standard forecasts of vehicles and traffic, *DOE Technical Memorandum H3/75*, HMSO, London, 1975.
Department of the Environment, *Transport Policy: A Consultation Document*, (2 vols.), HMSO, London, 1976.
Department of Motor Vehicles (California), Projected motor vehicle registration and drivers licences outstanding 1978–2005, Report No. 48, Sacramento, California, October 1978.

Department of Transport, *Report of the Advisory Committee on Trunk Road Assessment*, HMSO, London, 1978.

Department of Transport, National traffic forecasts, *DTp Interim Memorandum*, HMSO, London, 1978.

Department of Transport, *Forecasting Traffic on Trunk Roads: A Report on the Regional Highway Traffic Model Project*, HMSO, London, 1980.

Department of Transport, *National Road Traffic Forecasts*, HMSO, London, July 1980.

Hayfield, C.P., A method for forecasting car ownership in local areas using national travel survey data, *LTR1 Working Paper No. 12*, Department of Transport, London, 1979.

Teesside County Borough, *Teesside Survey and Plan*, HMSO, London, 1971.

Tulpule, A.H., Forecasts of vehicles and traffic in Great Britain, *Road Research Laboratory Report LR 288*, Road Research Laboratory, Crowthorne, 1969.

Tyne Wear Plan: A Transport Plan for the 1980's, A.M. Voorhees and Associates, Colin Buchanan and Partners, 1972.

Wootton, H.J., and Pick, G.W., A model for trips generated by households, *Journal of Transport Economics and Policy*, I, 1967, pp. 137–153.

Conclusion

IN OUTLINING some of the many problems associated with numerical prediction of car ownership, it has been possible to concentrate on the debate surrounding the different development procedures. It is evident that the development of valid methods has been hindered by failure to identify significant factors and variables, by data collection and availability, and general fallacies concerning the application of models based on non-human behavioural patterns in this particular area of consumer expenditure.

Widening the Base of Involvement in Forecasting

An additional problem is associated with the failure to consider the policies and interests of several powerful groups who, as well as possessing noteworthy views on future levels of ownership, could have a significant influence on those levels. Included in this category are a number of groups involved in the manufacture and use of vehicles, including user bodies such as the UK Automobile Association, local government, trades unions, insurance companies and, most important of all, car producers. With great levels of private and public investment in this sphere of manufacturing it is unlikely that the motor industry will not have some scientifically based policy on car ownership. Indeed, in the US, the major automobile producers employ large research teams, as well as independent consultants, in this very field. It goes without saying that the policy of manufacturers regarding supply of vehicles, price, variations in engine size and design features, marketing, and location of plant could have a significant effect on future patterns of ownership. A number of commentators have made observations on the manner in which private industry has influenced the nature of the nation's transport system, but such factors are not generally included in the discussion on modelling car ownership prediction. With so much economic importance now being placed on the motor industry and its ancillaries there would appear to be further scope for liaison between the industry and those working in the field of forecasting (particularly those in receipt of government sponsorship).

Government, on the other hand, should be fully aware of the extent to which the motor manufacturers and distributors can manipulate the new car market in such a fashion so as to ensure that the official predictions (upon which so much capital infrastructural investment is based) do ultimately prove accurate. In a similar fashion, governments can influence the level of demand for cars by direct intervention in the market (by taxation, statutory controls, and so on) as well as policy towards the provision of road space and public transport.

It is also evident that the academic base of the forecasting movement could be widened to produce a deeper insight into causal relationships and processes. Studies of the sociological and psychological aspects of car ownership are notable by their absence. Yet these factors can be expected to influence an individual's decision to purchase and use a private vehicle. Mogridge (1976) criticised the acceptance of the analogy between car ownership growth and other biological phenomena, on the basis that there was no empirical evidence for assuming that the ultimate pattern of ownership would follow such a similar path. This commentary has repeatedly noted the forecasting problems resulting from the special position which car ownership holds in western society, the fact that cars represent more than an alternative mode of personal transport, that they cannot be directly compared with other types of consumer good (capital cost alone places them in a special category), that they are subject to special forms of advertising, influencing and reflecting the power of social prestige and society values. Yet these factors form part of a wider social and economic systems, the workings of which ensure that demand for new cars continues to increase in an inelastic fashion until the saturation point is approached.

Continuing Criticism of Scientific Modelling

Although this book has concentrated on the detailed technical aspects of forecasting it should not be forgotten that there continues to be widespread scepticism regarding the basic philosophy of mathematical modelling. This is particularly noticeable in the general field of econometrics. The comments of Robbins (1978) are typical. In criticising the use of population and economic change as input parameters into car ownership models, he shared the concern of the Leitch Committee in the wide degree of variation in population forecasts since the 1930's, with the pessimism of that period being gradually replaced by "alarm" about a population explosion (in the 1960's forecasts) only to approach full circle in more recent projections. In the economic field, variations and errors have been compounded even further with the added problem of policy manipulation of data. It is unlikely that lay and professional scepticism, often resulting from misapplication of forecasts at second hand, will be reduced in the 1980's. Development of more sophisticated procedures is likely to be countered by greater uncertainty in future levels of key variables.

This problem is particularly acute in the field of econometric modelling. Following a decade of debate on the validity of different procedures, what is now emerging is a reappraisal of the simple time-series extrapolation approach used by the TRRL and UK Department of Transport. It is now clearly evident that causal models may appear more scientific and be less vulnerable to outside criticism, but they are heavily dependent upon accurate forecasting of at least one independent variable. In the case of income or wealth this introduces the additional complication of use of national projections based on objective reasoning or, alternatively, a policy figure, based on a government target of what the central administration "hopes to achieve". In the UK in particular, most administrations have over-predicted future growth in national wealth; often errors have been greater than 100 per cent, even over quite short time periods.

Added to this is the complex position which car ownership occupies within a wider social and economic framework, the fact that it is not a special variable capable of independent study but merely represents one of many factors included in a wider economic system, all of which are correlated to some degree. It could be argued that car ownership is a cause, as well as a consequence, of wealth production depending, of course, on the manner in which wealth is assessed.

The extrapolation trend technique, being primarily descriptive rather than analytical, by-passes some of these problems. It is assumed that the pattern of interaction between variables, which is evident from historical data, will continue in the medium term. Thus it has been argued that, since it involves only limited estimation of independent variables, it is more suitable as a basis for broad projections particularly at state or national level. This argument was further developed by Tanner (1979) in a call for each type of model to be examined on its appropriateness to the nature of the task to which it is applied. Thus the dichotomy which existed between basic UK models could be used to advantage. On the one hand, an unconditional general model could be used to produce simple forecasts, whereas a more sophisticated procedure would allow for alternative projections under "various alternative policy assumptions" as a key input into the decision making process.

New Considerations in Forecasting
Licence evasion and company fleets

Data availability and validity are representative of problems which are often overlooked in the search for superficial statistical significance. It is often assumed that official data from government and academic sources is both representative or all-embracing in unmodified form, without reference to the nature of its collection or subsequent transformation (many surveys involve some sampling procedure and do not involve entire populations). Thus, vehicle registration figures, for example, have been used as a standard input into many forms of model. Only recently Mogridge and others commented upon the high level of licence evasion in the UK.

The phenomenon of company ownership of fleets or subsidy is one which has received increased attention in recent times. Dix and Pollard (1980) have reiterated the growing importance of firms' cars as a factor in forecasting availability and use. In 1979, for instance, it was estimated that 43 per cent of new car and van registrations in Great Britain were company owned. More than 8 per cent of the total private vehicle stock fall into this category. Clearly, such proportions have significant consequences. National and individual income levels have been a consistent feature of modelling in the USA and UK, with a heavy emphasis on utility theory, and it is expected that this trend will continue. Clearly, the transfer of financial accountability from households to commercial bodies, either in the form of direct hiring, purchase or subsidy, requires further analysis, particularly as company cars tend to record high mileages (often at direct cost to the company, not the user).

Car use

Although the absolute number of vehicles owned may be of primary importance to car producers and designers of housing areas, the level of use of cars is of greater concern to a wide group of planners and engineers, often working in other transport fields (market evaluation) or estimating future demand for road space. Vehicle usage, a topic worthy of examination in its own right, has gained new importance in the context of uncertain fuel availability and cost.

Modal split modelling has traditionally received the same degree of attention as car ownership, often in joint form. The new emphasis on vehicle mileage has resulted in appreciation of the importance of driver behaviour. Level of use of vehicles has been shown to exhibit significant elasticity to fuel cost, at least in the short term, primarily as a result of the greater perceived importance of direct out-of-pocket expenditure and the limited availability of credit facilities compared to other forms of purchase (including cars).

Average mileage statistics require careful interpretation and tend to conceal significant changes in the pattern of vehicle use. This is particularly noteworthy in the increasingly important area of second and third car ownership at the household level. Because of the generally small number of multiple car owning households (at least in the UK), the use of such categories as an empirical basis for forecasting has been limited and subject to error. Yet it is the use of second cars which can have a significant influence, not only on overall average mileage, but the fortunes of other forms of passenger transport. The problem is particularly influenced by the growing availability and use of company owned vehicles as first cars.

Not surprisingly some of these phenomena are already the subject of research by the UK Department of Transport and other workers, as reported by Giles and Worsley (1979). In addition, the Department of Transport is actively engaged on examining other features of car ownership and use, namely the effect of household composition (a popular area of study in the US) and the effect of journey purpose on use. In contrast, in the US, there is heavy emphasis on future car size and fuel consumption, reflecting once more the different role played by the central administration and the automobile industry. What is clear, from this relatively brief commentary, is that the search for an acceptable solution of the "simple" problem of car ownership will continue for many years.

REFERENCES

Dix, M.C., and Pollard, H.R.T., Company financed motoring and its effect on household car use, *Traffic Engineering & Control*, 21, 1980, pp. 536–540.

Giles, G.E., and Worsley, T.E., Development of methods of forecasting car ownership and use, *Economic Trends*, No. 319, August 1979, pp. 99–108.

Mogridge, M.J.H., A personal view on the future of the car, *Traffic Engineering & Control*, 17, March 1976, pp. 101–104.

Robbins, J., Mathematical modelling — the error of our ways, *Traffic Engineering & Control*, 19, January 1978, pp. 32–38.

Tanner, J.C., Choice of model structure for car ownership forecasting, *Transport and Road Research Laboratory Supplementary Report SR 523*, Transport and Road Research Laboratory, Crowthorne, 1979.

Additional References

IN SPITE of its apparent simplicity the problem of car ownership forecasting has generated a significant amount of printed material. Several contributors have produced bibliographies, although none are widely available. Of these that of Fowkes and Button (1977 — revised 1979) is the most extensive. Several writers have consolidated earlier work in more recent summaries, a useful example being that produced by Tanner (1978). In addition, of course, many articles and books on general forecasting methods touch on the topic of car ownership, either directly, or by reference to widely used modelling techniques which have been applied in this specific field.

In view of the rapid nature of change in the developing of forecasting the following bibliography is mainly confined to works published since 1975 which have made a particularly significant contribution to the debate.

Armstrong, A.G., and Odling-Smee, J.C., The demand for new cars 1 — a theoretical model of replacement demand, *Oxford Bulletin of Economics and Statistics*, XL, 1978, pp. 281–302.

Bates, J.J., Gunn, H.F., and Roberts, M., A model of household car ownership (2 parts), *Traffic Engineering & Control*, 19, 1978, pp. 486–491, 522-566.

Button, K.J., Fowkes, A.S., and Pearman, A.D., Regional car ownership models: an investigation of functional form, Loughborough University, Department of Economics, Occasional Paper No. 19, 1977.

Button, K.J., Fowkes, A.S, and Pearman, A.D., Some outstanding problems in the causal modelling of car ownership at the local level, *Transportation Planning and Technology*, V, 1979, pp. 205–213.

Button, K.J., Fowkes, A.S., and Pearman, A.D., Car ownership in West Yorkshire: the influence of public transport accessibility, *Urban Studies*, XVII, 1980, pp. 211–216.

Dagenais, M.G., Application of a threshold regression model to household purchases of automobiles, *Review of Economics and Statistics*, August 1975, pp. 275–285.

Daly, A.J., Cost influences on car ownership and use. *PTRC Seminar Proceedings, Transportation Models*, PTRC, London, 1977.

Fowkes, A.S., and Button, K.J., A survey of techniques of car ownership forecasting, University of Leeds Institute for Transport Studies, Working Paper No. 92, 1977.

Fowkes, A.S., Button, K.J., and Pearman, A.D., Car ownership modelling and forecasting: an annotated bibliography, University of Leeds, Institute for Transport Studies, Working Paper No. 121, 1979.

Johnson, K., and Black, J., An aid to understanding car ownership — a study of households in four Melbourne suburbs, *International Journal of Transport Economics*, IV, 1977, pp. 148–167.

Lerman, S.R., Location, housing, automobile ownership and mode to work: a joint choice model, *Transportation Research Record*, No. 610, 1976, pp. 6–11.

Lerman, S.R., and Ben-Akiva, M., Disaggregate behavioural model of automobile ownership, *Transportation Research Record*, No. 569, 1975, pp. 34–55.

Mogridge, M.J.H., Traffic Forecasting, *Highways and Road Construction International*, 2 parts, part 1, XLV, No. 1811, 1977, pp. 5–9, part 2, XLV, No. 1812, 1977, pp. 5–9.

Mogridge, M.J.H., The effect of the oil crises on the growth in the ownership of and use of cars, *Transportation*, VII, 1978, pp. 54–65.

Pearman, A.D., and Button, K.J., Regional variations in car ownership, *Applied Economics*, VIII, 1976, pp. 231–233.

Smith, R.P., Consumer demand for cars in the USA. University of Cambridge, Department of Applied Economics, Occasional Paper 44, 1975.

Appendix 1

Summary of the TRRL Logistic Curve Procedure

THE METHODOLOGY which is outlined below was established by Tanner in the early 1960's and remained basically unaltered until the publication of TRRL Report LR 799 in 1977. At this time it was replaced by the power growth curve.

The logistic curve requires the estimation of three parameters in order that it may be fitted. These are:

- y — car ownership per person in year t, where $t = 0$ (at year zero)
- r_0 — rate of growth in car ownership at year zero (base year)
- s — the saturation level, which represents a future situation where new vehicle registrations are merely stock replacements, assuming a fairly stable household/population composition.

The logistic curve possesses a number of characteristics. One of these properties is that the rate of change in the variable (car ownership) is proportional to the product of the value of the variable in that year, and to the distance it is away from the saturation level. In the case of this curve:

$$\frac{dy}{dt} = ay_t(s - y_t) \tag{A1.1}$$

where a is a constant.

This differential equation is solved in the following form:

$$y = \frac{s}{1 + h\exp(-ast)} \tag{A1.2}$$

where h is a constant of integration.

At the base year ($t = 0$) the rate of change and rate of car ownership are as follows:

$$r_0 = a(s-y_0) \tag{A1.3}$$

and

$$y_0 = \frac{s}{1+h} \tag{A1.4}$$

Replacing a and h, the following equation for car ownership in future years is produced:

$$y_t = \frac{s}{1 + [(s-y_0)/y_0] \exp[-r_0 st/(s-y_0)]} \tag{A1.5}$$

The base year values of y_0 and r_0 are usually known (from survey information or government statistics). Before a forecast of car ownership in a future year (t) can be derived it is necessary to estimate s, the saturation level.

Estimation of the saturation level

Recently, the concept of a saturation level, and its means of calculation, have stimulated a great deal of discussion. The method outlined below, a basic extrapolation of a linear relationship between growth rate and rate of ownership, using local authorities as the spatial unit of measurement, remained central to the Tanner method for a number of years.

In straight statistical terms s is derived by regressing growth rate against car ownership rates which have been observed in a number of discrete areas (in most cases, UK counties). Estimates of the coefficients of the following regression equation are derived:

$$r = c + my + \epsilon \tag{A1.6}$$

where c and m are regression coefficients.

If r (rate of growth) = 0, then the corresponding value of y (saturation level) is given by:

$$s = \frac{-c}{m} \tag{A1.7}$$

The estimated value for s is then introduced into the earlier logistic curve equation to produce a value of y at a future design year t. Thus, an estimated saturation level of 0.4 cars per person, a base year value of car ownership of 0.128, and a base year growth rate of nine per cent would give a forecast of 0.256 for ten years hence and 0.348 ten years later.

Modifications to allow for variations in income and motoring costs

In 1974 Tanner introduced a refinement on the basic method to allow for the effect of variations in income per head and motoring costs. The logistic curve equation was modified thus:

$$y_t = \frac{s}{1 + [(s-y_0)/y_0][i_t/i_0]^{-bs}[p_t/p_0]^{-cs}\exp[-as(t-t_0)]} \quad (A1.8)$$

where
    y = car ownership in year t
    s = saturation level
    y_0 = car ownership at base year
    t = year
    i = income per head (at fixed prices)
    p = cost of motoring (at fixed prices), and
    a, b and c are constants.

This method formed the basis for the associated Department of Transport forecasts of 1975.

Appendix 2

Population Projection Methods used in the UK

CAR OWNERSHIP rates form only part of the input into the general forecasting procedure. In the UK a primary variable influencing the total number of vehicles in existence, and corresponding traffic flows, is the size and structure of the national population. Although this book has highlighted the likely sources of error associated with car ownership forecasting models, variation in estimates of total vehicle stock are clearly dependent upon levels of population and affluence. Traffic levels are further influenced by land use distribution, economic growth, fuel prices and national or regional transport policy. Forecasting the future pattern of all these factors is a complex business and falls beyond the scope of this work — suffice it to say that most of these variables have been subjected to a level of academic attention similar to that afforded to car ownership. Such is the intensity of pressure for development of land in the UK that in the field of land use distribution conventional forecasting has been replaced by a system of strictly controlled allocation.

National population forecasts have been produced under the auspices of the Registrars General since the 1920's, and on an annual basis since 1955. These are rightly described as projections, for they are derived by use of a complex extrapolatory procedure incorporating past trends in birth and death rates, female fertility and migration. No direct allowance is made for future variations in national wealth although all the four variables noted above are, to an extent, proxies for income. In contrast to the recent methods used in national car ownership and traffic forecasts, a single projection has been produced each year. Although the 1980 forecasts include the effects of several alternative scenarios, it is expected that a single middle projection will be used to avoid inconsistencies between government departments and other users. The inflexibility resulting from this is countered to an extent by the issue of new projections each year, although, in practice, little change is involved.

APPENDIX 2   147

The procedure involves a careful analysis of birth and death rates, fertility, and migration as the exclusive components influencing population change. Fertility is dependent upon marital condition which itself is a reflection on rates of divorce, spinsterhood and remarriage.

The most recent projections involve the use of mortality rates based on group experience since 1945. It is assumed that rates in various age groups will decrease in geometrical progression during the projection period (40 years). Estimation of migration is more liable to error as a result of a complex pattern in flow in and out of the country since the Second World War. It is accepted that this component, more than any other, is subject to sudden changes related to economic conditions, and government policy, at home and abroad, but that the traditional net outflow will continue. Fertility rates have fluctuated widely during the present century and although the reasons for past changes may be clear in hindsight they are often either so intangible or related to major global events as to be almost impossible to forecast. This is evident from the various official forecasts of future birth rates. In 1970 it was estimated that the number of live births in 1978 would be approximately 880,000. In fact the real figure was 596,000. The total period fertility rate had reduced from 2.41 in 1970 to 1.75 eight years later. Introducing the concept of "variant projections" into the problem, a "high" fertility assumption could see the number of live births almost double that for an alternative "low" rate.

The component method uses several mathematical formulae to produce forecasts on the basis of past trends and assumptions on future patterns. The most important of these is used to estimate the number and age of persons in a given year in the following form:

$$^{n}P_{x} = {}^{n-1}P_{x-1}(1 - {}^{n-1}q_{x-1/2}) \pm {}^{n}M_{x} \qquad (A2.1)$$

where $^{n}P_{x}$ = number of persons at mid-year n aged x last birthday

$^{n}M_{x}$ = net number of migrants in the period mid-year $(n-1)$ to mid-year n who would be aged x last birthday at mid-year n

$^{n-1}q_{x-1/2}$ = probability of death within a year for a person aged $(x-1)$ last birthday at mid-year $(n-1)$.

Clearly the use of such a model involves the risk of error. Much depends upon the validity of projecting past trends in components and, among other considerations, the time range of the data. Although it could be argued that the absolute number of vehicles in existence and their level of use is more related to overall rates of income, residential density, and so on, the actual number and structure of households is likely to be one of many demographic factors influencing future patterns.

REFERENCE

Office of Population Censuses and Surveys, *Population Projections*, Series PP2, No. 10, London, HMSO, 1980.

Appendix 3

Car Ownership in the United States and Great Britain 1900–1979

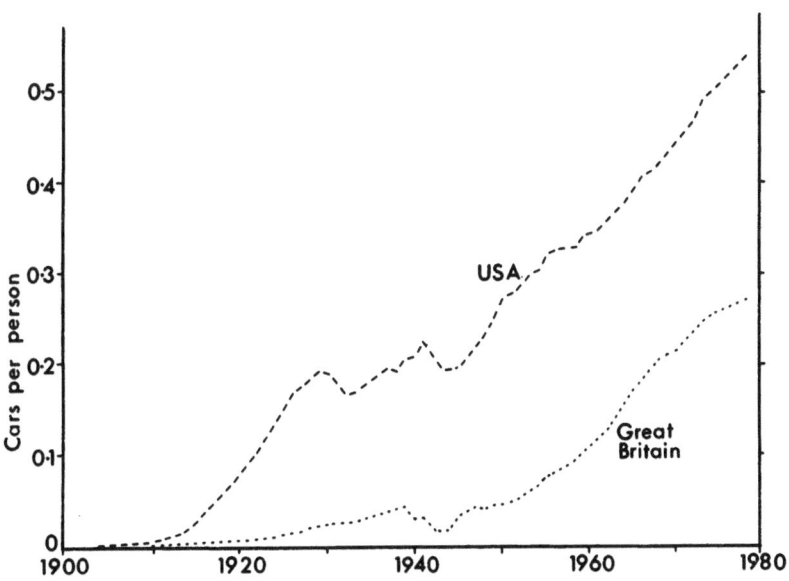

Index

Accessibility cycle   106
Accessibility indices   38, 57
Adams, J.   19
   criticism of Tanner and DOE
     forecasting methods
     48–50, 56
Adelaide   115, 118
Administrative boundaries
   effect on density rates   37
Advertising of new cars   136
Advisory Committee on Trunk Road
   Assessment (ACTRA)   53–56,
   132
Aggregate-demand model   97
Analogy between car ownership and
   biological phenomena   136
Analogy with consumer durables   7,
   14–16, 49, 91, 136
Atkinson, L.J.   91
Australia
   forecasting car ownership in
   114–118
Auto-correlation   104
Automobile Association (AA)   135
Automobile choice model   110
Auto sales   98, 100
Auto stocks   98
Ayres, R.   101

Bandeen, R.A.   92
Bates, J.J.   56
   criticism of Mogridge model
     36–37
   criticism of new TRRL
     model   67

   development of extrapolatory
     model   37
   modification of model by
     Fairhurst   38
Beesely, M.E.   24–26, 97
Behavioural models
   in Australia   118
   in the USA   106–111
Ben-Akiva, M.E.   60, 95, 106
   criticism of Wharton model
     101–102
   joint choice model   108,
     110–111
Bendigo   115–116
Bennett, R.J.   53
Bennett, W.B.   104
Biological diffusion analogy with car
   ownership   50, 55
Bone, A.J.   88
Bonus, H.   49
Bottiny, W.H.   97
Brain, R.   116
Briefs, G.E.   92
Brighton   135
Brisbane   115
Buchanan Report   17, 24
Buckingham County Structure Plan
   125
Bunch, H.M.   102
Burns, L.D.   105–106
Business-owned cars   31, 72
Button, K.H.   35, 37, 40–41, 49–50,
   69–71, 143, 141
Buxton, M.J.   40

151

California
    Department of Motor Vehicles
        local forecasts 133
        evidence of saturation in car
            ownership 83
Campbell, M.E. 81–83, 97
Canberra 114–115, 118
Car ownership/income probability
    curves 34, 35, 100, 126
Car ownership rates in the USA and
    Great Britain 149
"Car ownership time" 70
Car price index 58
Car producers 135
"Car purchasing income" 35–36, 73,
    92
Car use 138–139
Category analysis 31–35, 88
    use by Cleveland County
        Council 132
    use by Lancashire County
        Council 126–127
Causal models
    in the UK 6, 29, 56–61
    in the USA 96–114
Census of Population (1971) 66
*Central and North Lancashire
    Structure Plan* 126–127
Central government involvement 3,
    4, 9–13, 44–79
Chandler, K.N. 14
Chase model of demand
    variables used 98–100
Chicago Area Transportation Study
    (CATS) 85–88
Chow, G.C. 92
Cleveland County (UK) 129–132
COBA 44, 47, 55, 72
Commonwealth Bureau of Roads
    (Australia) 116
Company cars 138
Conjoint analysis 111
Consultation document on transport
    policy 125

Consumer demand for cars 94
County forecasts 26, 47
Cragg, J.G. 104
Cramer, J.S. 15
Credit terms 98
Cumulative log-normal curve 31, 49

Dawson, R.F.F. 15
Department of the Environment
    (UK) 44
    national forecasts 47–52
Department of Motor Vehicles
    (California)
    local forecasts 133
Department of Transport (UK) 122,
    124, 137
    1978 traffic forecasts 62,
        64–66, 75
    1980 traffic forecasts 74–77
Detroit 106
Deutschman, H.D. 94–95
Devon County Structure Plan 125
Disaggregate model 70–73
    in USA 103–112
    use in Telford New Town
        71–72
    virtues recognised by
        ACTRA 55
Distance from CBD 97
Distance from London 42
District councils (England and
    Wales) 123
Dix, M.C. 138
Downes, J. 72–73
Downs, A. 102
Driving licences
    forecasting future rates 76
Duesenberry, J.S. 16
Dunphy, R.T. 105
Durham County Structure Plan 125
Dyckman, T. 97

Econometric models 29–43, 94
    examined by ACTRA 55
    in the USA 90–92

INDEX   153

Economic activity rates   42, 67
Economic growth   68
Economic Man   111
Eggert, R.J.   103, 105
Elasticity   91–92, 98, 106
Eldridge, D.A.   35
Energy consumption   5
Energy crisis   7, 12, 45, 52, 73
Engine capacity   73–74
Evans, A.W.   31
Evans, M.   98
Expenditure on cars   4
Extrapolation
    in Australia   115–117
    in the USA   95–96
    reappraisal of technique
       137–138

Factors influencing car ownership   8, 10
Fairhurst, M.H.   38
Family Expenditure Survey   58–59
Farrell, M.J.   15, 91, 103
Fauth, G.R.   106
Fishwick, F.   41
Forecasting models
    type of   5–6
Fowkes, A.S.   35, 37, 49–50, 69, 70–71, 141
Frechtling, J.A.   92
Fuel efficiency and consumption   110–111
Future Automobile Population Stochastic Model (FAPS)   102–103

Gamma distribution   60
Geology and car ownership   40
Giles, G.E.   139
Golob, T.F.   105–106
Golomb, D.H.   102
Goss, A.   13
Greater London Council (GLC)   35, 124
Gross domestic product   22

Growth in car ownership   2, 149

Haggett, P.   53
Hayfield, C.P.   132
Hendriks, G.   103
Hereford County Structure Plan   124
Herrmann, P.G.   17, 24, 26, 47, 70
Hobart   115
Household based models
    in USA   103–112
Household structure   33, 56, 103, 139
Hoxie, P.   105
Huang, D.S.   103
Hutchinson, M.J.   118
Hymans, S.   98

Income distribution
    future estimation of   60
Income levels
    effect on car ownership   29, 31, 42, 57, 66, 72, 81, 85, 88–89, 91, 97, 129–130, 145
Inflation   45
Insurance companies   135
Ioannou, A.C.   50

Janosi, P.E.de   103
Johnson, T.   106, 110
Joint choice model   108

Kain, J.F.   24–26, 97, 106
Kanwit, E.L.   97
Kirby, H.R.   50–52, 70
Kulash, D.J.   100

Lansing, J.B.   97, 103
Latitude and car ownership   40
Launceston (Australia)   115
Leathers, N.T.   97
Leeds
    forecasts for   24–25
Leicester County Structure Plan   125
Leitch Committee   53–56, 136–137
Lerman, S.R.   106–109

Level of education and car ownership 103
Licence evasion 73, 138
Licensed drivers 112
Local forecasting 22–23, 122–132
    in the USA 133
Local plans 123
Logistic curve 16, 45–46, 55, 76, 90, 143–145
    used in Australia 115–117
    used in Cleveland (UK) 129–132
    used in Tyne and Wear 128
Logit models 106, 109
Love, C. 110

Manski, C.F. 110
Martin, B.V. 88
McCracken, P.W. 103, 105
Melbourne 115–116
Memmott, F.W. 88
Ministry of Transport forecasts 1945–59 13
Mogridge, M.J.H. 136, 56, 73
    early modelling 29–31
    estimation of saturation 31–32, 35
    GLC model 35–36
Motor trade 5
Motoring costs 145
Motorways 13
Mueller, E. 97
Mullen, P. 71–72
Multiple-car ownership 60, 139
    in the USA 112–113
Mundarri, D.H. 105

Naive models 6, 25
National forecasts
    for Great Britain 62, 64–66, 74–77, 125–127, 132
    for USA 83
National Traffic Model 56, 122
National Travel Survey (NTS) 37, 132

Nerlove, M. 92
New South Wales 116
Nicolaidis, G.C. 105
Non-registration of vehicles 73
North East Lancashire Structure Plan 126–127
North Yorkshire County Structure Plan 125
Northampton County Structure Plan 125

Office of Population Censuses and Surveys (OPCS) 148
Oi, W.Y. 88

Pearman, A.D. 69, 71
Perception of cost 98, 111
Perth (Australia) 115–116
Pick, G.W. 32–35, 126
Planning loops 53, 68, 87
Plowden, W. 9–13
Policy dependence and car ownership 68
Policy scenarios 109
Pollard, H.R.T. 138
Pooled model 71
Population density 26, 31, 41, 63, 66, 71, 92, 96
Population forecasts 22–23, 146–147
Population growth 67
Population structure 67
Power growth curve 62–63
Probability model of auto choice 110
PTE's 122–123
Public housing design 13
Public transport and car ownership 4, 26, 33, 38, 72, 105–106
Public transport use 38

Quasi-logistic curve 31

Rateable values and car ownership 41
Rates of car ownership
    in Great Britain 3

INDEX   155

in the U.S. 112–113
Regional Highway Traffic Model
    (RHTM) 56–59, 122
    compatability with TRRL
        model 76–77
    criticism by Tanner 60–61
    limitations of car ownership
        model 59–61
    modifications to the model
        75–77
Regional variations in car
    ownership 11, 36, 38–42, 70
Registrars General's population
    forecasts 146–147
Regression
    use in Australia 115–118
    use in California 133
    use in demand model 9
    use in Teesside 129–130
    use in TRRL model 144
    use to calculate saturation
        19–21, 45, 49, 71
Residential density
    influence on car ownership 24,
        37, 41, 97, 99, 100, 104,
        116
Residential location
    influence on car ownership 24,
        37
Reynolds, D.J. 14
Road Construction Units 122
Road Research Laboratory (RRL)
    13, 15, 17–28, 44
    criticism of RRL model 24–27
Robbins, J. 136
Roberts, M. 56
Roos, C.F. 91
Rudd, E. 14

Saturation levels
    criticism of Tanner method
        26, 45, 48
    early US estimates 81, 83, 90
    estimates in Britain in 1920's
        11, 12
    estimation by Mogridge 31–32

    estimation using Tanner
        method 19, 21, 63–64,
        144
    evidence of in 1950's 14
    evidence of in USA 2, 97, 112
    forecasts for Australian cities
        115–117
    Kirby's observations 50–52
    regional variation 22–24
    use in RHTM 59–60
Saunders, M.B. 73
Scenario tree 102–103
Schmidt, R.E. 81–83, 97
Scrappage rates 100–101
Second-hand cars 15
Sherman, L. 110
Shuldiner, P.W.
Sleeman, J.F. 39–40
Smith, J.E.R. 73
Social and economic planning 5
Society of Motor Manufacturers and
    Traders 11
Solo, D. 92, 98
Spatial variation in car ownership 70,
    92
Special Roads Act 13
Stock adjustment models 92, 98
Structure plans 123–131
Suits, D. 92, 98
Sydney Area Transportation Study
    (SATS) 116–118
von Szeliski, V. 91

Tanner, J.C.
    accuracy of forecasts 45
    basic logistic model 143–145
    bibliography 141
    examination of US trends
        95–96
    forecasts of 1958 14
    new model 61–69
    observations on choice of
        model 69
    observations on RHTM 60–61
    reply to critics of 1974
        forecasts 48–49

research at RRL 17–28
reservations on forecasts 22, 64, 67
retention of saturation level 46
Taste factors 91
Taxation on car sales 4
Teesside Structure Plan 131
Teesside Survey and Plan 129
Television sets 15, 49
Telford new town 71
Tetlow, J. 13
Time series extrapolation 14, 18
Todd, T.R. 83–85
Toowoomba 115
Townsville 115
Trades unions 135
Traffic Engineering Handbook 80–81
Traffic in Towns 17, 24, 26
Train, K. 95, 106, 109–110
Transit scenarios 109
Transport and Road Research Laboratory (TRRL) 45, 55, 59, 61–65, 69, 76–77, 137, 143–145

Transport Planning Associates (TPA) 35
Transport Policies and Programme (TPP) 123
Trip generation models 33
Tulpule, A.H. 20–21, 27, 48
Tyne and Wear Structure Plan 129

Uhler, R.S. 104
United States Department of Transportation forecasts 112
Urban development 26, 97, 114
Utility theory 29, 138

West Germany
    logistic growth in car ownership 49
West Midlands 33
West Riding of Yorkshire 41
White, M. 71–72
White paper on transport policy 52–53
de Wolff, P. 14, 90
Wootton, H.J. 32–35, 126
Wyckoff, F.C. 98

For Product Safety Concerns and Information please contact our EU
representative GPSR@taylorandfrancis.com
Taylor & Francis Verlag GmbH, Kaufingerstraße 24, 80331 München, Germany

www.ingramcontent.com/pod-product-compliance
Lightning Source LLC
Chambersburg PA
CBHW070618300426
44113CB00010B/1575